Stuff That Works!

A TECHNOLOGY CURRICULUM FOR THE ELEMENTARY GRADES

Mechanisms & Other Systems

Gary Benenson and James L. Neujahr
Project Directors, City Technology

Heinemann
Portsmouth, NH

Heinemann
A division of Reed Elsevier Inc.
361 Hanover Street
Portsmouth, NH 03801–3912
www.heinemann.com

Offices and agents throughout the world

Stuff That Works!
City College of New York
140 Street & Convent Avenue, Room T233,
New York, New York 10031
(212) 650-8389 tel.; (212) 650-8013 fax
citytechnology@ccny.cuny.edu

This project was supported, in part, by the
National Science Foundation
Opinions expressed are those of the authors and not necessarily those of the Foundation

Project Staff
Gary Benenson, *Project Director, City College of New York, School of Engineering*
James L. Neujahr, *Project Co-Director, City College of New York, School of Education*
Dorothy Bennett, *Education Development Center/Center for Children and Technology*
Terri Meade, *Education Development Center/Center for Children and Technology*

Advisory Board
William Barowy, *Lesley College*
David Chapin, *City University of New York*
Alan Feigenberg, *City College of New York*
Ed Goldman, *Brooklyn Technical High School*
Patricia Hutchinson, *The College of New Jersey*
Neville Parker, *City College of New York*
Peter Sellwood, *Consultant, United Kingdom*
Ron Todd, *The College of New Jersey*

Production Staff
Gary Benenson, *General Editor and Lead Author*
Lorin Driggs, *Editor*
Doris Halle Design NYC, *Design and Graphics*
Maria Politarhos, *Photography*

Teacher Associates/Coauthors
Katherine Aguiar, *CES 42, Bronx, NY*
Helen deCandido, *Retired*
Mary Flores, *CES 42, Bronx, NY*
Angel Gonzalez, *Family Academy, New York, NY*
Sandra Jenoure, *PS 155, New York, NY*
Shirley Monterroso-Nieves, *PS 96, New York, NY*
Annette Purnell, *CES 42, Bronx, NY*
Minerva Rivera, *Harbor Academy, New York, NY*
Verona Williams, *PS 60, Bronx, NY*

Library of Congress Cataloging-in-Publication Data
ISBN 0-325-00468-4
CIP data is on file with the Library of Congress.

Printed in the United States of America on acid-free paper

06 05 04 03 02 VP 1 2 3 4 5

CONTENTS

FOREWORD

In a world increasingly dependent on technology—where new ideas and tools pervade our personal and civic lives and where important choices hinge on our knowledge of how things and people work—the imperative that all students should learn to understand and use technology well should be obvious. Yet in the American curriculum, still overstuffed with tradition and trivia, there is little room in the day for learning and teaching about important ideas from technology and very few resources for educators who want to engage their students in learning for the 21st century.

Stuff That Works! is a groundbreaking curriculum. It provides a set of carefully chosen and designed activities that will engage elementary students with the core ideas and processes of technology (or engineering, if you prefer). Elementary school is the ideal place to begin learning about technology. It is a time in students' development when they are ready and eager to take on concrete rather than abstract ideas. The concepts and skills presented in ***Stuff That Works!*** will support more advanced learning in mathematics, science, and technology as students move up through the grades.

But there is much more to ***Stuff That Works!*** than a set of activities. As a matter of fact, the activities make up less than a third of the pages. ***Stuff That Works!*** also includes helpful resources for the teacher such as clear discussions of the important ideas and skills from technology that their students should be learning; stories of how the materials have been used in real classrooms; suggestions for outside reading; guidance for assessing how well their students are doing; and tips on implementation. I hope teachers will take time to make full use of these valuable resources as they use ***Stuff That Works!*** If they do, they can help their students take the first, critical steps towards technological literacy and success in and beyond school.

George D. Nelson, Director
American Association for
the Advancement of Science (AAAS)
Project 2061

INTRODUCTION

What Is Technology?

Stuff That Works! Mechanisms and Other Systems will introduce you to a novel and very engaging approach to the study of technology at the elementary school level. In education today, the word *technology* is most often associated with learning how to use computers, and that is certainly important. But learning how to use a particular kind of technology is not the same thing as learning how and why the technology works. Children learn about computers as *users* rather than as students of how computers work or of how to design them. In fact, computer analysis and design require technical knowledge that is beyond most adults, let alone elementary-aged children. Fortunately, there are many other examples of technology that are much more accessible than computers and that present many of the same issues as computers and other "high-tech" devices.

The purpose of technology is to solve practical problems by means of devices, systems, procedures, and environments that improve people's lives in one way or another. Understood this way, a computer is no more an example of technology than...

- the cardboard box it was shipped in,
- the arrangement of the computer and its peripherals on the table,
- the symbol next to the printer's ON/OFF switch,
- or the ballpoint pen the printer replaces as a writing device.

A box, a plan for the use of table space, an ON/OFF symbol, and a pen are examples of technologies you and your students will explore in this and the other *Stuff That Works!* guides.

The *Stuff That Works!* approach is based on artifacts and systems that are all around us and available for free or at very low cost. You need not be a technical guru or rich in resources to engage yourself and your students in technology. The topics of *Mechanisms and Other Systems* include devices you can find in your kitchen, closet or bathroom, such as eggbeaters, nail clippers and umbrellas; as well as electrical appliances, bicycles, faucets, and mousetraps.

Why Study Technology in Elementary School?

Below is a graphic summary of the process of "doing" technology as we present it in this book. The study of technology challenges students to identify and solve problems, build understanding, develop and apply competence and knowledge in a variety of processes and content areas, including science, mathematics, language arts, and social interaction.

The teachers who field-tested these materials underscored that these activities helped their students to:

- observe and describe phenomena in detail;
- explore real objects and situations by creating models and other representations;
- identify salient aspects of problems;
- solve authentic problems;
- use evidence-based reasoning;
- apply the scientific method;
- ask thoughtful questions;
- communicate in oral, written, and graphic form;
- collaborate effectively with others.

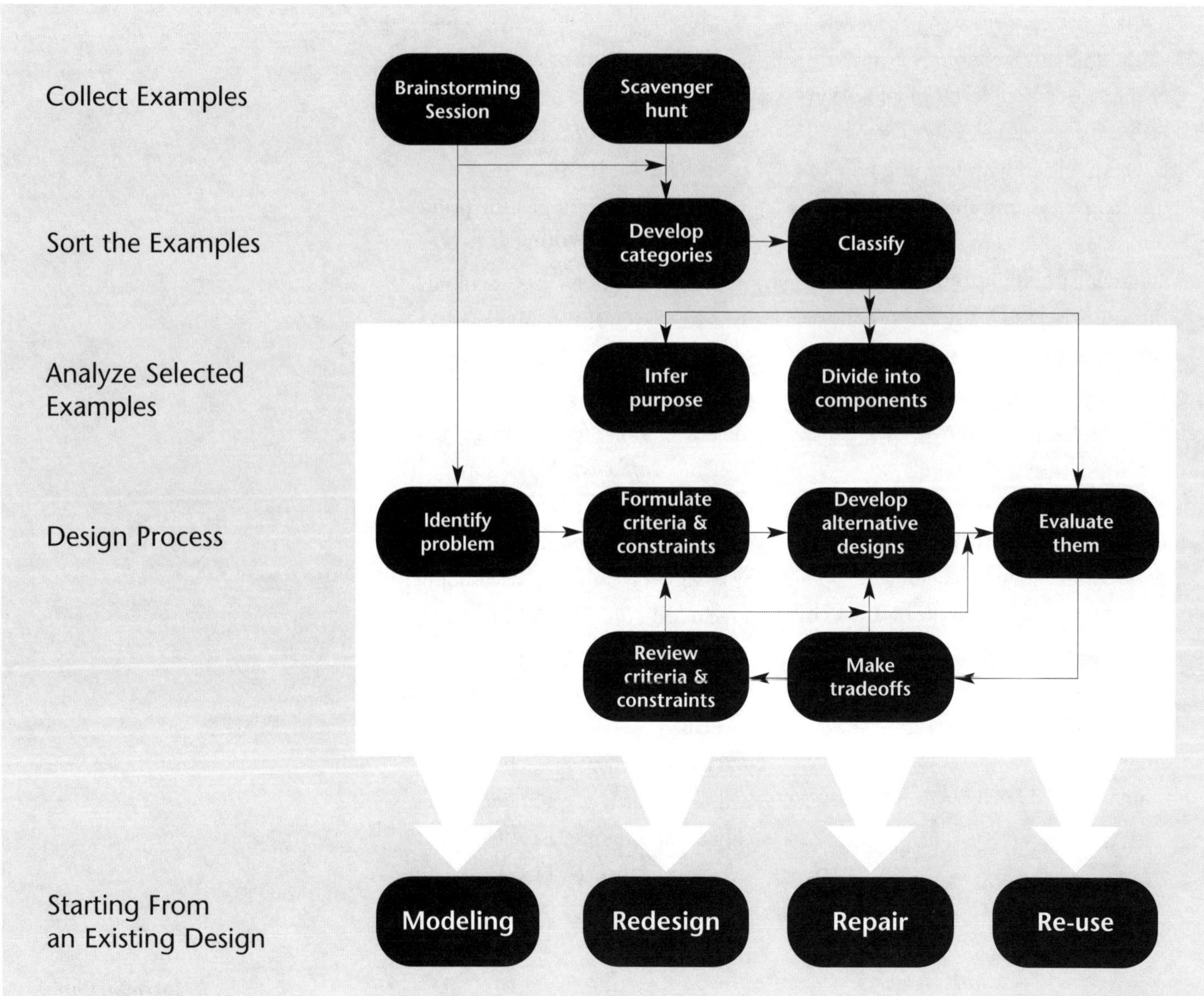

Educational Goals for *Mechanisms and Other Systems*

Mechanisms and Other Systems deals with devices and systems that transform motion, convert energy, and/or process information to get a job done. The content and activities presented here will help you meet these instructional goals:

- Introduce and explore fundamental themes of systems, inputs and outputs, cause-and-effect, models;
- Illustrate and explore concepts of force, distance, motion, lever, simple machine, friction, electric current, electric circuit, information, control, feedback and energy;
- Demystify common artifacts, and by extension, technology in general;
- Promote literacy as students formulate problems and find effective ways to communicate with others in order to achieve and document solutions;
- Develop process skills in observation, classification, generalization, use of materials, modeling, and design;
- Provide rich opportunities for group work.

How This Guide Is Organized

This *Stuff That Works!* guide is organized into the following chapters.

Chapter 1. *Appetizers* suggests some things you can do for yourself, to become familiar with the topic. You can do these activities at home, using only found materials. They will help you to recognize some of the technology that is all around you, and offer ways of making sense of it.

Chapter 2. *Concepts* develops the main ideas that can be taught for and through the topic. These include ideas from science, math, social studies and art, as well as technology. It also reviews what is known from relevant cognitive research.

Chapter 3. *Activities* contains a variety of classroom projects and units related to the topic, including those referred to in Chapter 4. Each activity includes prerequisites, goals, skills and concepts; materials, references to standards and teacher tips; and sample worksheets.

Chapter 4. *Stories* presents teachers' narratives about what happened in their own classrooms. Their accounts include photos, samples of children's work and children's dialog. Commentary by project staff connects the teachers' accounts with the concepts developed in Chapter 2.

Chapter 5. *Resources* provides a framework supporting the implementation of the activities. It includes an annotated bibliography of children's literature and a discussion of assessment opportunities.

Chapter 6. *Standards* shows how the activities and ideas in this book address national standards in technology, science, math, English language arts (ELA), and social studies.

How to Use This Guide

Different teachers will come to this book with different needs and objectives. However, regardless of your background, instructional approach, and curricular goals, *we strongly recommend that you begin with Chapter 1, Appetizers.* There is simply no better way to become acquainted with a topic and to understand what your students will be facing than to try out some of the ideas and activities for yourself. Chapter 1 guides you through that process.

The content and approach presented in *Mechanisms and Other Systems* are based on the premise that processes of design are central to the practice of technology, just as inquiry is the central activity of science. While no two design problems are the same, there are some features that characterize any design task:

- It should solve a problem of some sort.
- It must have more than one possible solution.
- There must be an effort to test the design.

A problem is like a trigger that initiates a design process. Often the problem is not well-formulated, a vague kind of "wouldn't it be nice if…" In making the problem more specific, it is often helpful to list some criteria the design must address. In trying to satisfy these criteria, the designer is never completely free to do whatever he or she wants. There are always constraints, which could involve cost, safety, ease of use, and a host of other considerations.

Mechanisms and Other Systems presents a number of activities that include elements of design, but are not full-scale design projects. These elements of design are modeling, redesign, repair, and re-use.

- **Modeling** requires both a very close look at the original design and its modification to incorporate the use of different materials. A lot can be learned by observing how the substitution of materials affects the operation of the model. Modeling is only one kind of design activity that starts with an existing solution.
- **Redesign** starts with an existing but inadequate design. It involves analyzing the weaknesses of the original design and then figuring out how to correct them.
- **Repair** is a variant of redesign. It takes place after the existing design has already failed. Redesign and repair projects often use new materials or techniques to accomplish the original purpose.
- **Re-use** is a complementary kind of design activity where the original materials are used for a new purpose.

The concepts of redesign, repair, and re-use are of particular importance in a society that has been widely criticized for its wasteful practices. These three concepts are considerably more accessible than the more widely advanced notion of recycling, whose full implementation requires expensive equipment and specialized technical knowledge.

There is no one way to do design. It is a non-linear, messy process that typically begins with very incomplete information. Additional criteria become apparent as the design is implemented and tested. New constraints appear that were not originally evident. It is often necessary to backtrack and revise the original specifications. Such a messy process may seem contrary to the work you usually expect to see happening in your classroom. However, we encourage you to embrace the messiness! It will justify itself by improving students' competence in reasoning, problem-solving, and ability to communicate not only what they are doing but also why they are doing it and what results they expect.

A Brief History of *Stuff That Works!*

The guides in the *Stuff That Works!* series were developed through collaboration among three different kinds of educators:

- Two college professors, one from the School of Education of City College of the City University of New York, and the other from the City College School of Engineering;
- Two educational researchers from the Center for Children and Technology of the Education Development Center (CCT/EDC);
- Thirty New York City elementary educators who work in the South Bronx, Harlem, and Washington Heights.

This last group included science specialists, early childhood educators, special education teachers, a math specialist, a language arts specialist, and regular classroom teachers from grades pre-K through six. In teaching experience, they ranged from first-year teachers to veterans with more than 20 years in the classroom.

During the 1997-98 and 1998-99 academic years, the teachers participated in workshops that engaged them in sample activities and also provided opportunities for sharing and discussion of classroom experiences. The workshop activities then became the basis for classroom projects. The teachers were encouraged to modify the workshop activities and extend them in accordance with their own teaching situations, their ideas, and their children's interests.

The teachers, project staff, and the research team collaborated to develop a format for documenting classroom outcomes in the form of portfolios. These portfolios included the following items:

- lesson worksheets describing the activities and units implemented in the classroom, including materials used, teacher tips and strategies, and assessment methods;
- narrative descriptions of what actually happened in the classroom;
- samples of students' work, including writing, maps and drawings, and dialogue; and
- the teachers' own reflections on the activities.

The lesson worksheets became the basis for the **Activities** (Chapter 3) of each guide. The narratives, samples of student work, and teacher reflections formed the core of the **Stories** (Chapter 4). At the end of the two years of curriculum development and pilot testing, the project produced five guides in draft form.

During the 1999-2000 academic year, the five draft guides were field-tested at five sites, including two in New York City, one suburban New York site, and one each in Michigan and Nevada. To prepare for the field tests, two staff developers from each site attended a one-week summer institute to familiarize themselves with the guides and engage in sample workshop activities. During the subsequent academic year, the staff developers carried out workshops at their home sites to introduce the guides to teachers in their regions. These workshops lasted from two to three hours per topic. From among the workshop participants, the staff developers recruited teachers to field-test the *Stuff That Works!* activities in their own classrooms and to evaluate the guides. Data from these field tests then became the basis for major revisions that are reflected in the current versions of all five guides.

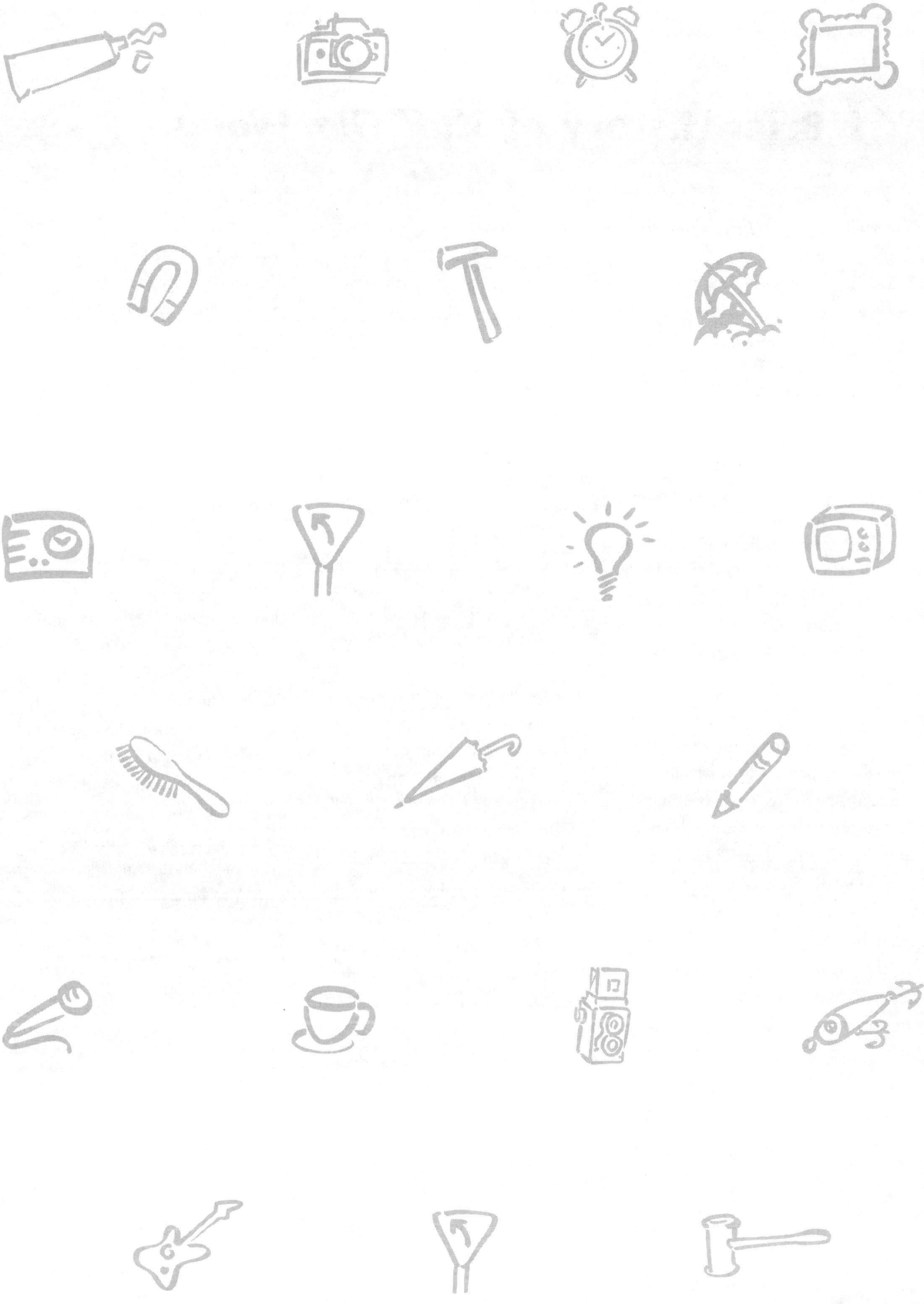

Chapter 1

Appetizers

This chapter will help you get started by suggesting ways you can learn about mechanisms and other systems for yourself. Some of the ideas will also work with your students later on, but for now the focus is on you. By trying out some of the activities in this chapter, you will become much more familiar with the topic and see its potential for your classroom.

Watch a carousel. What makes the horses go up and down? Look closely at a toddler's pull-toy. As you pull it horizontally, parts of it move up-and-down and maybe even sideways. How does this happen? Open an umbrella. When you release the catch, how does this allow the ring to slide up the shaft? As you move the ring up, how does it make the skin of the umbrella stretch outward and open?

Mechanisms All Around

The carousel, pull-toy, and umbrella are all examples of *mechanisms,* or devices with moving parts. A good way to prepare yourself for teaching mechanisms is to look for mechanisms in your own life. Think of it as a scavenger hunt. Here's a list of common mechanisms you have a good chance of finding at home or at school. Later in the chapter, we'll take a close look at some of these devices and suggest some simple ways to analyze how they work.

- **Kitchen mechanisms:** arm-operated corkscrew, rotary can opener, jar opener, nutcracker, pedal-operated wastebasket, egg beater, ice-cream scoop, toaster, tea-kettle spout, coffee maker, garlic press, egg topper, pizza tray holder, garlic peeler, jar opener, crank-operated peeler, juicer, salad spinner, mixer, blender
- **Bathroom mechanisms:** nail clipper, eyelash curler, tweezers, retractable mirror, scale, electric razor, hair dryer, spray cleaner, toilet, faucet, adjustable shower head
- **Classroom mechanisms:** hole puncher, scissors, pencil sharpener, retractable ballpoint pen, bulldog clip, clamp-on binder, ring-binder opener, clipboard
- **Office mechanisms:** adjustable desk lamp, stapler, staple remover, self-inking rubber stamp, typewriter, postal scale, copy machine, fax machine
- **Fold-up mechanisms**: folding chair, folding table, ironing board, baby carriage, shopping cart, umbrella, luggage cart, clothes drying rack, sewing box with expandable trays, foldable collator

- **Yard mechanisms:** lawnmower, lawn sprinkler, pruning shears, gate latch, hose spray attachment, snow thrower, hedge clipper
- **Appliance mechanisms:** VCR, tape player, sewing machine, watch, clock, camera, rotary-dial phone, record player, oscillating fan, retractable-cord vacuum cleaner, steam iron with sprayer
- **Computer mechanisms:** printer, scanner, CD ROM drive, disk drive, mouse, trackball
- **Bicycle mechanisms:** pedal, chain-and-sprockets, handbrake, gearshift, derailleur, pump, bell, speedometer
- **Door and window mechanisms:** deadbolt lock, automatic door closer, casement window crank opener, venetian blind, window shade
- **Toy mechanisms:** windup toy, pull toy, pop-up, yo-yo, Rubik's cube, Jacob's ladder, toy car, transformer, flexible action figure, water gun

Obviously, there are many more possibilities. Look around!
How many more items can you add to the list?

1-1: **Pedal-operated wastebasket**

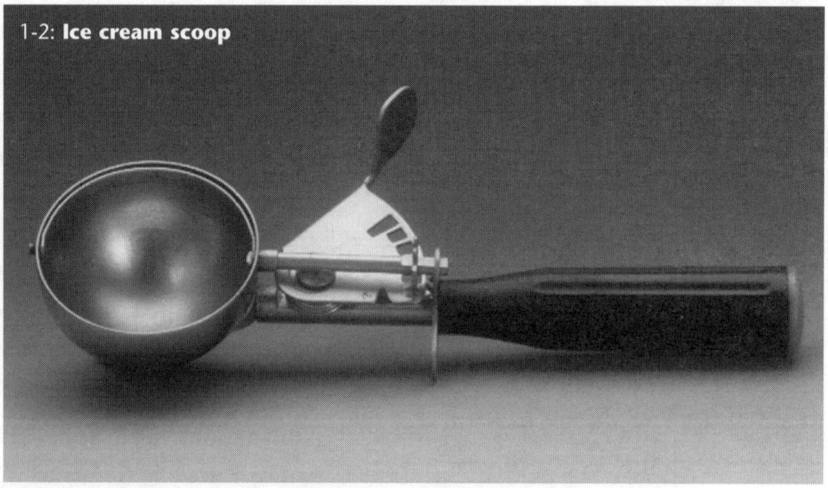
1-2: **Ice cream scoop**

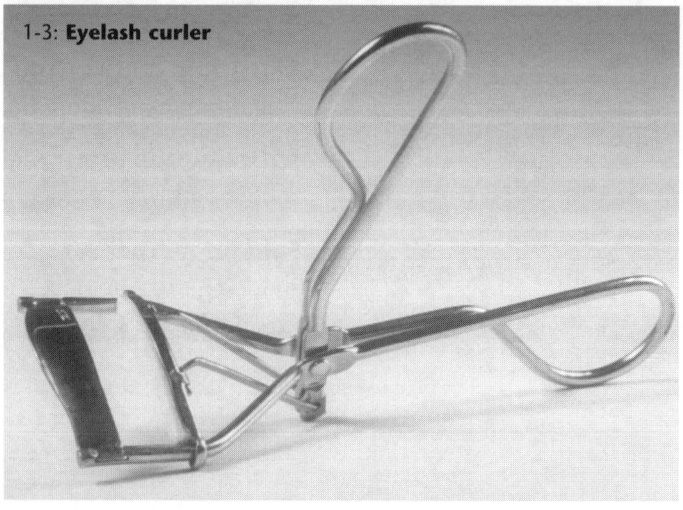
1-3: **Eyelash curler**

1-4: **Pencil sharpener**

1-5: **Foldable collator**

1-6: **Umbrella**

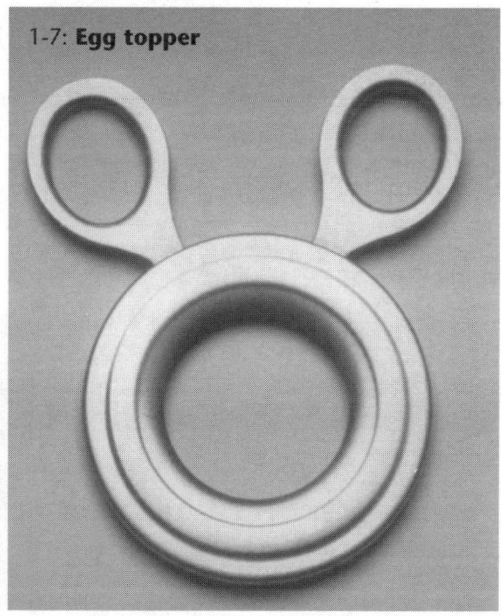

1-7: **Egg topper**

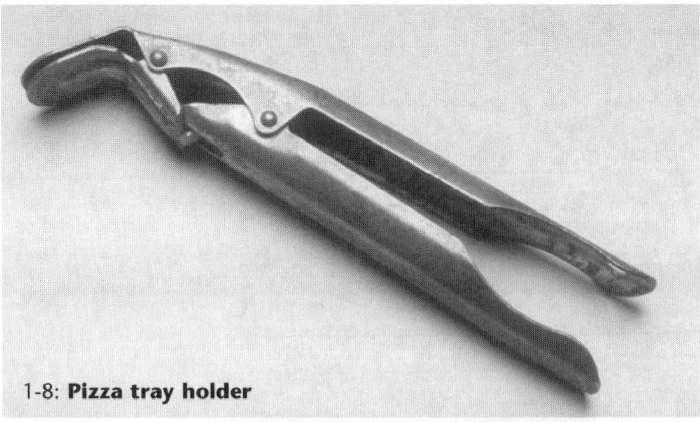

1-8: **Pizza tray holder**

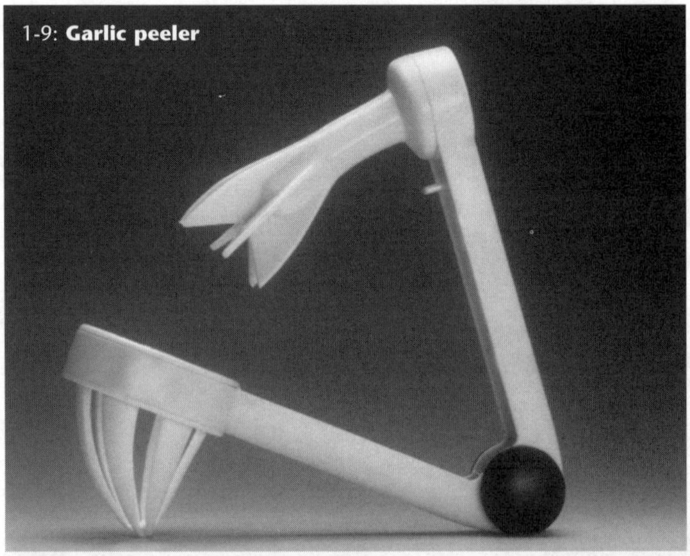

1-9: **Garlic peeler**

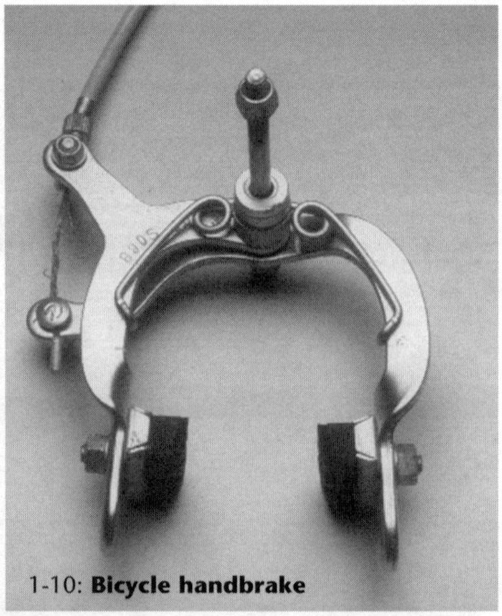

1-10: **Bicycle handbrake**

1-11: **Wind-up toy**

Tips for Mechanism Collectors

You can find a lot of simple mechanisms just by looking around at home and at school. But don't stop there.

The trash is an excellent source for appliances and computer equipment. This stuff is free, it's often in good mechanical condition (though obsolete or missing parts), and you don't have to worry about breaking it when you take it apart to see how it works. Schools throw out obsolete office and computer equipment from time to time. Ask the school custodian where large items are stored before being disposed of. Try to arrange to have him or her notify you when an item of interest is thrown out.

Other good sources are the basements of apartment buildings, shopping malls, community centers, churches and synagogues. Most custodial staff people are more than happy to cooperate if you explain your needs to them.

Look around your neighborhood and town. Many towns and cities have bulk trash pick-up days when people are allowed to put larger items on the curbside. It may be worth spending a few dollars if you spot a hard-to-find item of particular interest at a tag sale, street fair, or flea market. Those are good places to get typewriters, old cameras, rotary-dial phones, and phonographs without spending a lot of money.

Removing just a handful of screws or prying off outer plates and housings is usually all it takes to expose the mechanism. That often means the device's working days are over. Before you take anything apart, make sure no one cares about it or intends to use it again.

To find out how the mechanisms in complicated devices work, you have to take them apart. For that you'll need a few screwdrivers. If you don't already own some, pick up both a set of full-size flat and Phillips screwdrivers and a pocket set of precision screwdrivers, again including both standard and Phillips varieties. The full-size screwdrivers come in handy for taking apart larger items like printers, sewing machines, and typewriters; the little ones are good for taking apart cameras, wind-up toys, cassette tape players, and the like.

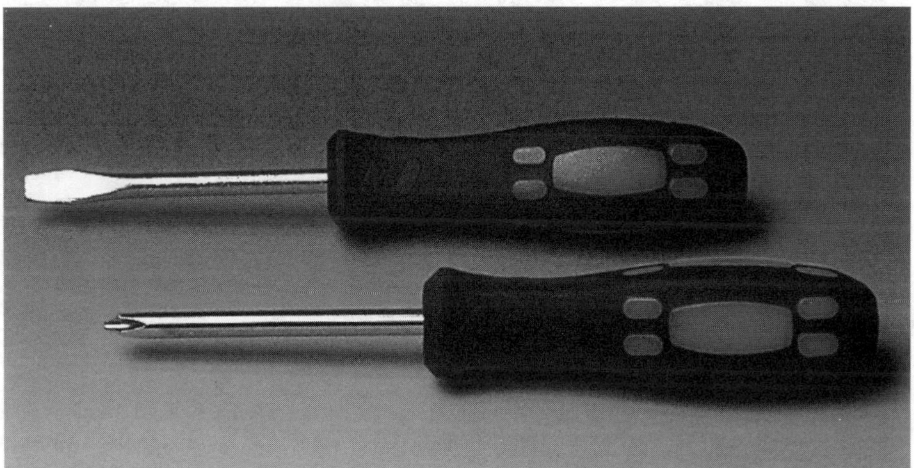

1-12: **Flat head** (top) **and Phillips head** (bottom) **screwdrivers**

Sorting Your Collection

A useful and enjoyable way to begin making sense of your collection is to think about different ways of sorting the mechanisms you have found. When children engage in sorting activities, they are developing and exercising some of the most basic science process skills: exploring similarities and differences, defining categories, and classifying. There is also a practical reason for sorting: if you have to store your mechanisms for later use, classifying them first will make it much easier to find the ones you want.

How to establish categories for sorting depends on the nature of your collection and also on some basic philosophical decisions. Before reading further, think about how you would answer these questions: What principles should determine categories for sorting? Should mechanisms be understood in terms of what they do or in terms of their most basic components? The first method adopts a *technology* perspective, because it focuses on the purpose of the device, from the user's point of view. The second approach is typical of science; it analyzes the device down to its fundamental parts.

Table 1-1

SIX WAYS OF SORTING MECHANISMS

Principle	Typical Categories	Comments
By type of user, place of use	Toy, kitchen utensils, bathroom, outdoors, male/female, architectural hardware, etc.	This was the principle used in categorizing the brainstorming list above.
By overall function	Devices that cut, devices that squeeze, devices that grab, devices that fasten, etc.	These are categories that describe the purpose of the device as a whole. The mechanisms are seen as a means to an end.
By degree of complexity	Simple levers, compound levers, more than one type of simple machine, requires one or two hands to use, etc.	This is an easy way to start, because it involves less background knowledge than most of the other methods.
By type of motion	Parts that move in a straight line (linearly), parts that rotate, parts that oscillate (rotate back and forth)	These catgories are based on the geomentry of a mechanism. They require you to trace the moving parts, and understand the paths they follow (see next chapter).
By type of machine	First, second and third class levers, inclined plane, pulley, gear, screw, wheel & axle.	Here are the classic "science categories" which require you to reduce a mechanism to its simplest component parts (see next chapter).
By purpose of each element of the mechanism	Change the location of motion, change the type of motion (e.g., from linear to rotary), amplify force, amplify range of motion	These are more technical categories, which reflect an analysis of why the designer chose a particular mechanism in each case (see next chapter).

Table 1-1 suggests half a dozen ways of sorting mechanisms. The first three require little or no background knowledge, while the last three depend on some analysis of mechanism types and motions. These topics are covered in detail in the next chapter.

Looking Closely at the Mechanisms

Now it's time to focus on how the mechanisms in your collection work. We suggest you begin with a fairly simple mechanism, such as an eyelash curler. If you don't already have one, you can buy one at a drugstore for about $2. Move the handles back and forth a few times to get a sense of how the curler works. Look closely at the two "extreme positions"—when the curler is fully open and when it is completely closed. These positions are illustrated in Figures 1-13 and 1-14.

Take a closer look at the mechanism. Hold one handle in one hand. Move the other handle with the other hand. Notice that the V-shaped arm (marked by arrow) also moves, pushing the lower grip toward the fixed upper one. When you use the device this way, the handle you move is the *input,* and the grip that closes is the *output.*

The ice-cream scoop is another simple mechanism. Figures 1-15 and 1-16 show the scoop in its open and closed positions, respectively. The thumb lever, which is the input, is at the end of a triangular plate with a series of slots cut into it. When you push on the lever, these slots turn a little gear, which is attached to the C-shaped slicer (see arrow) by a shaft. The slicer is the part that cuts a scoop of ice cream away from the bowl to release it. Can you find the spring that returns the lever and slicer to their original positions when you remove your thumb?

1-13: **Eyelash curler, open position**

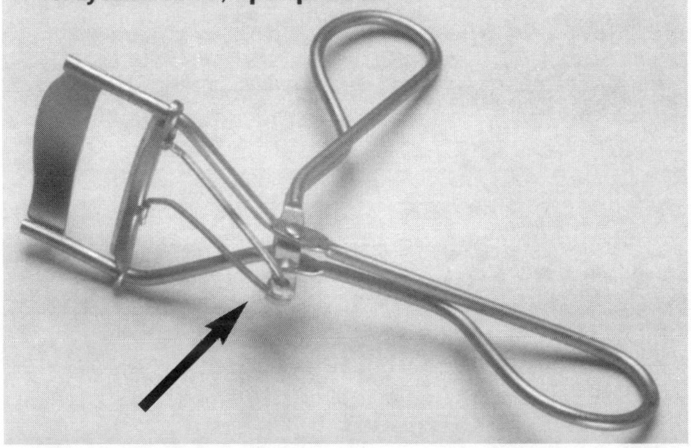

1-14: **Eyelash curler, closed position**

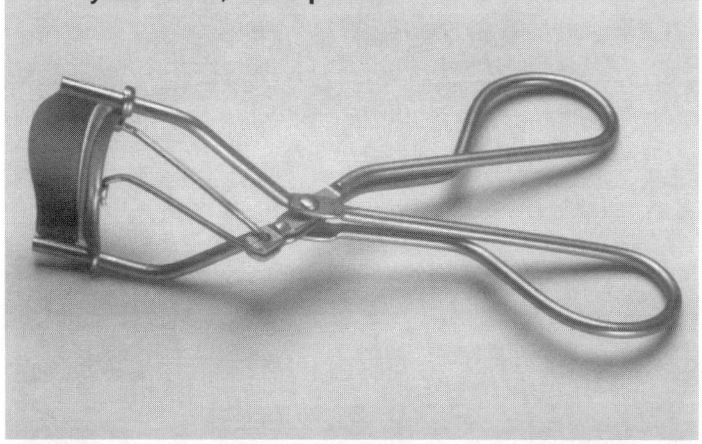

1-15: **Ice cream scoop, open position**

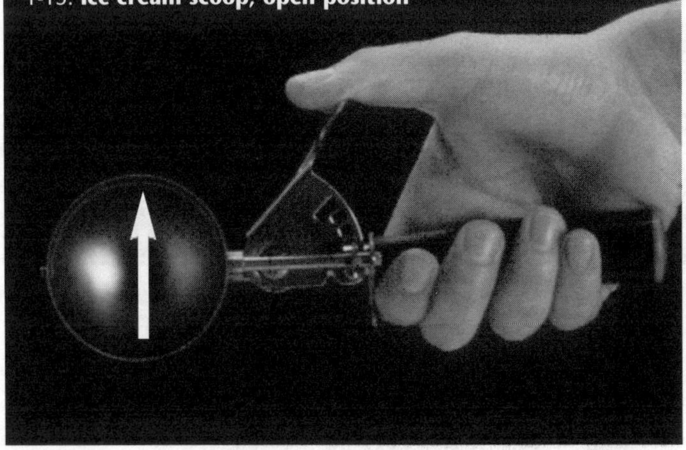

1-16: **Ice cream scoop, closed position**

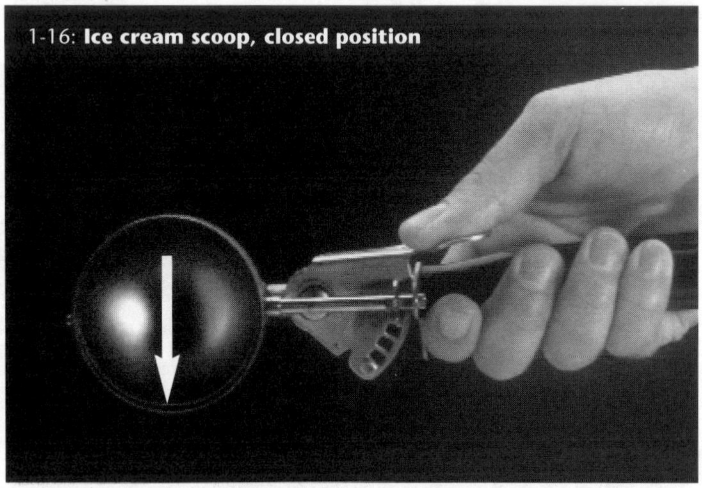

Many mechanisms are much too complicated to analyze completely, but not too hard to figure out in little pieces. Figure 1-17 shows a tape cassette player. When the "EJECT' button is pressed, a little door pops open, releasing the cassette. What makes it do so? By removing a few screws, it is not hard to find out. Figure 1-18 shows the same tape player with the cover removed, nearly ready to be examined to see how the "EJECT" mechanism works. In Figure 1-19, you can see the main body, removed from the case. The "EJECT button is at the top left. When this button is pressed, as in Figure 1-20, the long rod (see arrow) just under the frame moves to the right, forcing up the latch that opens the door holding the cassette.

1-17: **Cassette tape player with finger on the "EJECT" button**

1-18: **Cassette tape player, partly disassembled**

1-19: **Inside of cassette tape player, door in "closed" position**

1-20: **Same view as 1-19, door opened by "EJECT" button**

What's Inside of Me?

Hidden Mechanisms

Here are some challenges presented in increasing order of difficulty. The inputs and outputs in the following devices are obvious, but the mechanisms are concealed. Can you imagine and sketch what's going on inside of them? The answers are presented later. Right now, your challenge is to try to figure these out on your own.

Challenge #1: The Deadbolt Cylinder Lock

The common deadbolt lock shown in Figure 1-21 poses a little problem. Turning the handle or the key causes the bolt to slide up or down inside the two catches on the left. What happens inside to convert a turning motion to a straight-line motion? What holds the mechanism in the locked and unlocked positions until the key or the knob is turned?

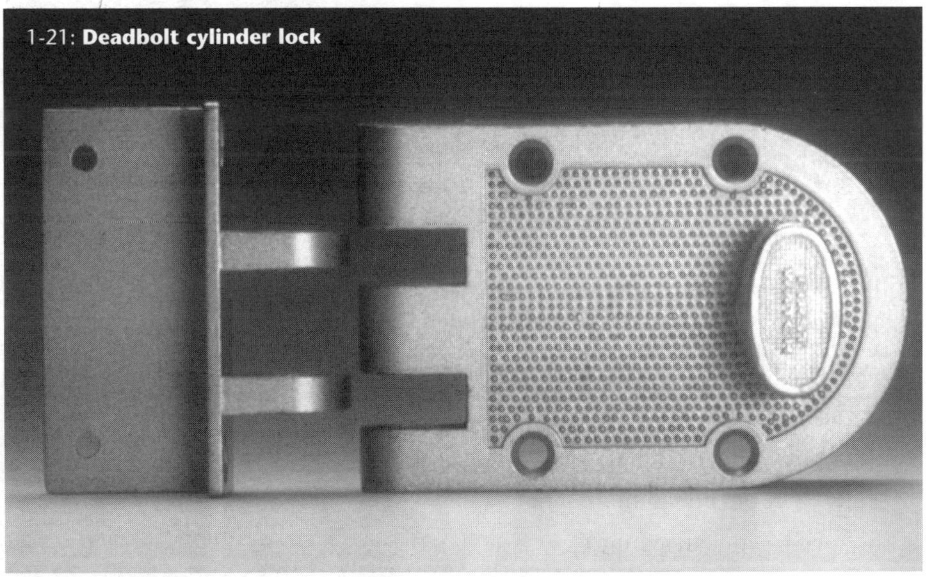

1-21: **Deadbolt cylinder lock**

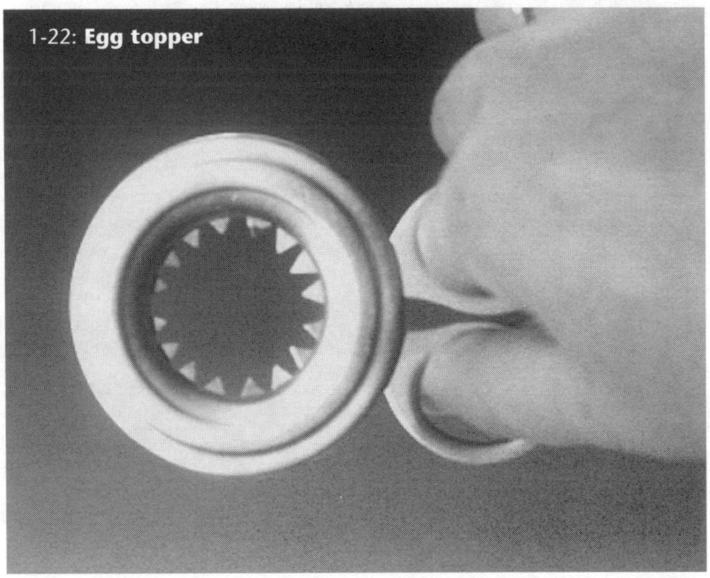

1-22: **Egg topper**

Challenge #2: The Egg Topper

The next example is the egg topper originally shown in the open position in Figure 1-7. This is a device for cracking the shell of a soft-boiled egg, so the top can be removed. Its two handles, when they are squeezed together, force the teeth all around the top of the egg. (See Figure 1-22.) What makes the teeth move toward each other when the handles are squeezed? What makes them return to their concealed positions when the handles are released?

Challenge #3: A Wind-up Toy

The wind-up toy crab shown in Figure 1-11 walks sideways, arms waving and eyes bobbing, when the key in the back (not shown) is wound up and released. It walks by alternating between two positions, which are shown in Figures 1-23 and 1-24, respectively. In Figure 1-23, the crab's legs are retracted and the body is resting on the ground. The arms are out and the eyes are up. As the spring unwinds, the legs come down, the arms go up, and the eyes come down, as shown in Figure 1-24. Then it repeats the same cycle, over and over again. While the legs are resting on the ground, the body shifts slightly to the right. When the body reaches the ground again the next time, it has actually walked about a quarter of an inch.

How does the unwinding motion of the spring get changed into the various motions of the legs, arms and eyes, and how does the body manage to shift sideways while the legs are down? We'll show you the answer to this one, too.

1-23: **Wind-up crab with body on the floor**

1-24: **Wind-up crab with legs on the floor**

Hidden Mechanisms Revealed: What's Inside

Challenge #1:
The Deadbolt Cylinder Lock

The lock can be taken apart by removing a few screws. It can be a little tricky to put back together, however, because one of the springs may fly off as the cover is removed. The mechanism is shown in the open position in Figure 1-25A. The knob or key (not shown) operates the two ears, which turn in the direction shown by the arrows. As they do so, they push against the movable plate carrying the two bars up and down. To close the lock, you have to turn the knob clockwise (in this view) so the ears turn through the positions shown in Figures 1-25B, C, and D. As the bottom ear in Figure 1-25A turns to the left, it pushes the plate up, lifting the two bars, which first become visible in Figure 1-25C. When the lock is turned the other way, the opposite ear has the job of pushing the plate down, disengaging the two bars.

1-25A, 1-25B, 1-25C, 1-25D: **Deadbolt cylinder lock in four positions starting with "open" (A) and proceeding to "closed" (D).**

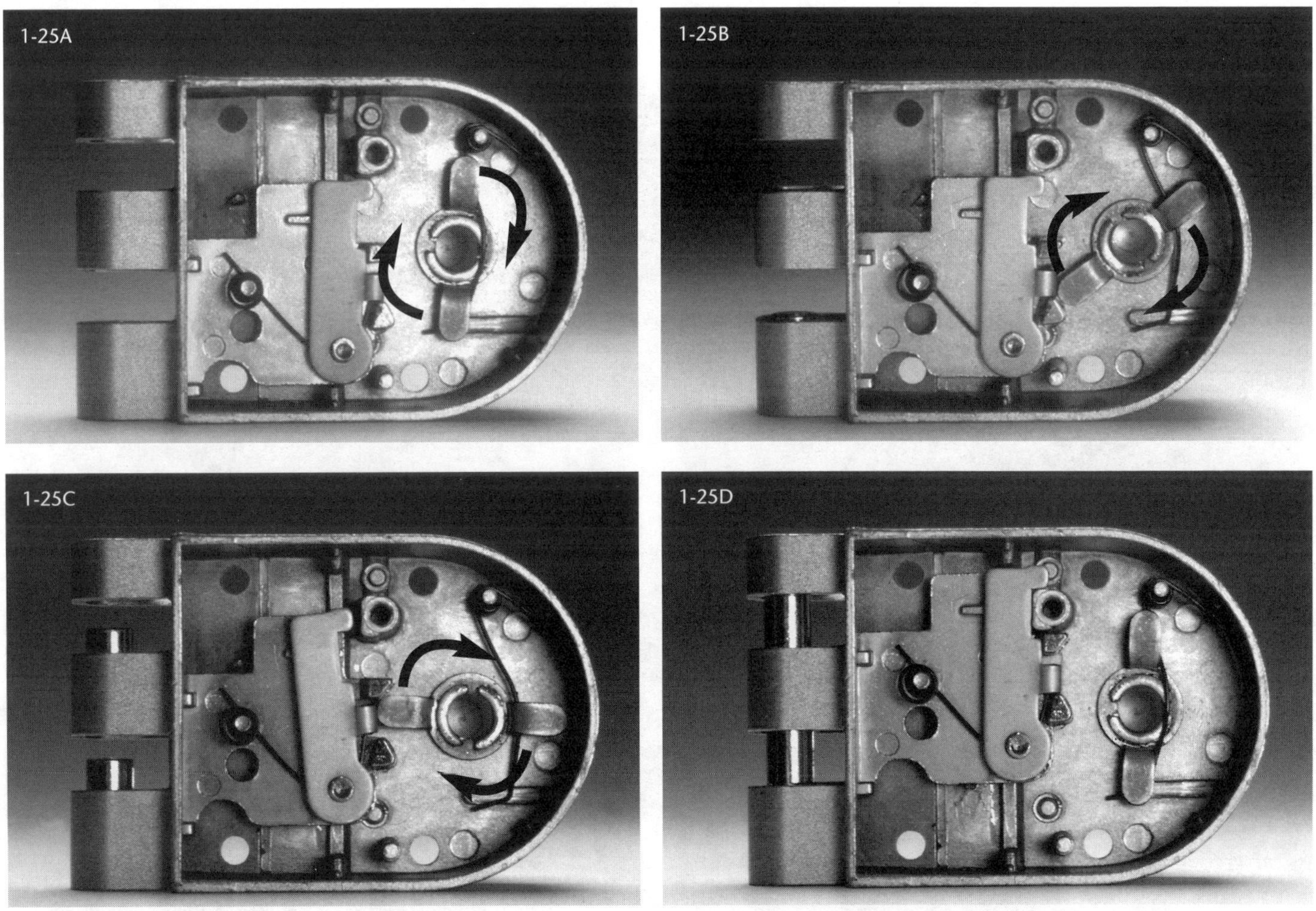

Challenge #2: The Egg Topper

Figure 1-26 shows the inside of the egg topper. The left side has been removed and the right side has its teeth retracted. Each side has two sets of teeth. The pivots are arranged so that moving a handle toward the center forces both sets of teeth inward. (See Figure 1-27.)

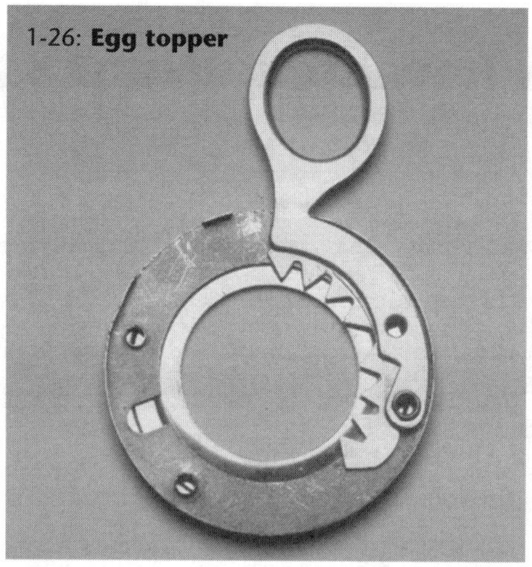

1-26: **Egg topper**

1-27: **Egg topper**

Challenge #3: Wind-up Toy Crab

The inside of the toy is shown in Figure 1-28 in the "body down" position, and in Figure 1-29 in the "legs down" configuration. Figure 1-30 shows the gearbox, which uses the energy of the wound-up spring to turn the little white wheel with the off-center peg. As the wheel turns, the peg alternately lifts the legs and eyes (Figure 1-28) and lowers them (Figure 1-29) by forcing the frame up and down. In Figure 1-31, the frame has been removed, revealing how each arm is mounted. As the frame moves the end of the arm, it makes the entire arm rotate around the pivot near the center (marked by arrow). Because of this, when the frame goes up, the arms go down, and vice-versa.

1-28: **Wind-up crab with legs and eyes up**

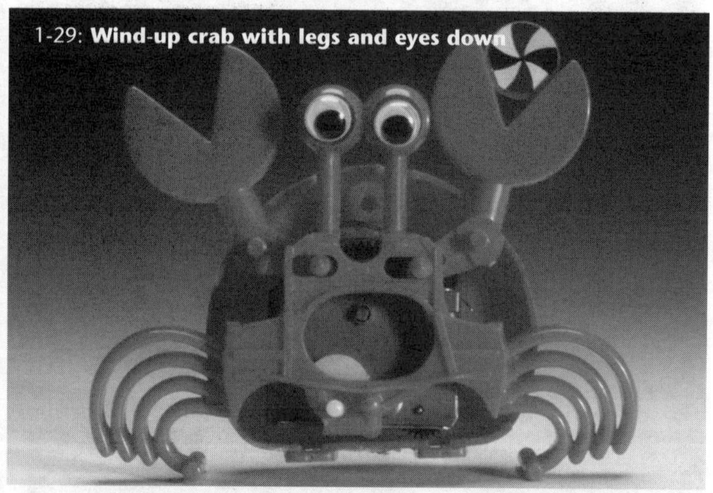

1-29: **Wind-up crab with legs and eyes down**

1-30: **Gearbox with off-center peg which operates all parts of wind-up crab**

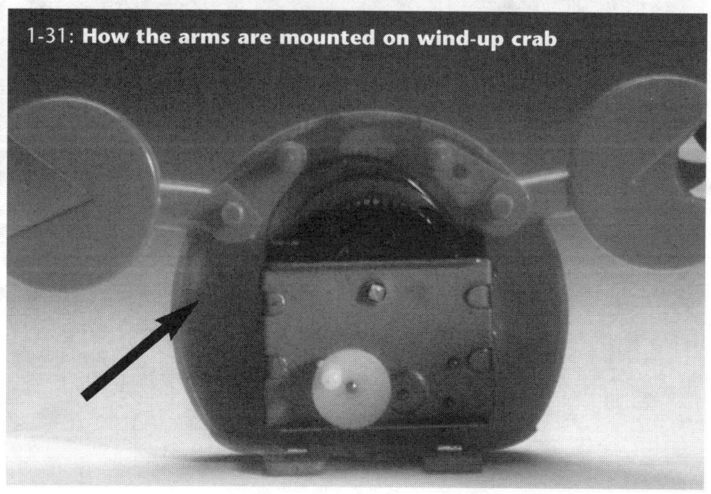

1-31: **How the arms are mounted on wind-up crab**

Making Models

A great way to get the "feel" for how a mechanism works is to make a model of it. Modeling is an activity that bridges the gap between analysis and design. In order to make a model of something, you have to analyze it carefully, identifying input and output, and all of the parts in between. In many mechanisms, the shapes of the parts and exact locations of the pivots are critical to its operation. A model that is not done to scale may not work the same way as the original.

Modeling involves many aspects of design. The modeler has to decide what size to make it, what materials to use, and so on. Once you have made the model, you still have to test it to see if it works properly, just as you would with any "from-scratch" design problem. Because there are so many factors to consider, it is likely that you will need to redesign some aspect of your model, and test it again!

1-32: **Folding chair**

The easiest mechanisms to model are two-dimensional. Two-dimensional *linkages* consist of three or more *links* (which are rigid rods) joined by either *pivots* or *sliders* or both. A pivot allows one link or arm to rotate around another, while a slider allows a link to move in a straight line. These mechanisms are called two-dimensional because all of their links have to move within the same plane surface.

Some other mechanisms, such as foldable baby strollers and automatic sponge mops, are more complicated because motion can take place in all three dimensions at once.

In selecting mechanisms for modeling, it is better to begin with the simpler two-dimensional linkages. Some examples of two-dimensional linkages are the eyelash curler (Figures 1-3, 1-13, 1-14), collator (Figure 1-5), pizza tray holder (Figure 1-8), bicycle handbrake (Figure 1-10), folding chair (Figure 1-32) and retaining ring pliers (Figure 1-38). Other examples from the scavenger hunt list at the beginning of this chapter include the pedal-operated wastebasket, adjustable desk lamp, ironing board, drying rack, and the expandable tray sewing box (or tool kit or fishing tackle box).

All of the devices listed above are actually three-dimensional, but the basic mechanism operates in two dimensions and can be modeled on a flat surface. An example is the folding chair (Figure 1-32), whose mechanism can be modeled using flat pieces of cardboard.

Figure 1-33 shows a two-dimensional model of one side of the chair. It incorporates the entire fold-up mechanism of the actual chair. When some teachers in a workshop were trying to model this folding chair, they had difficulty placing the notch correctly. As a result, they couldn't get the seat to be exactly horizontal in the open position. (See Figure 1-34.) These problems led to several cycles of design and redesign until they finally achieved a model like Figure 1-33.

1-33: **Diagram of folding chair**

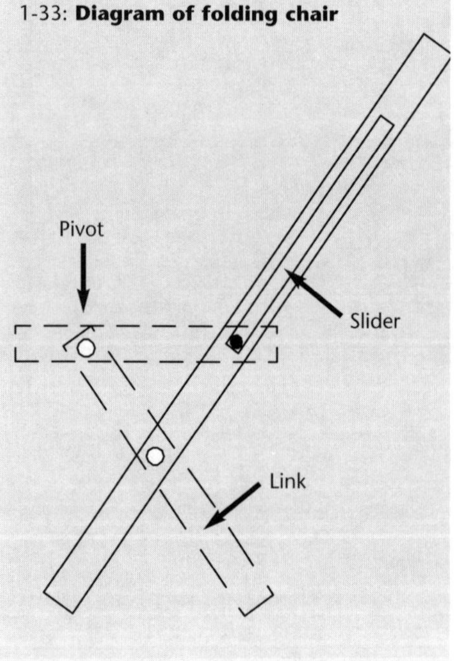

1-34: **Folding chair models with notch too high (left) and too low (right)**

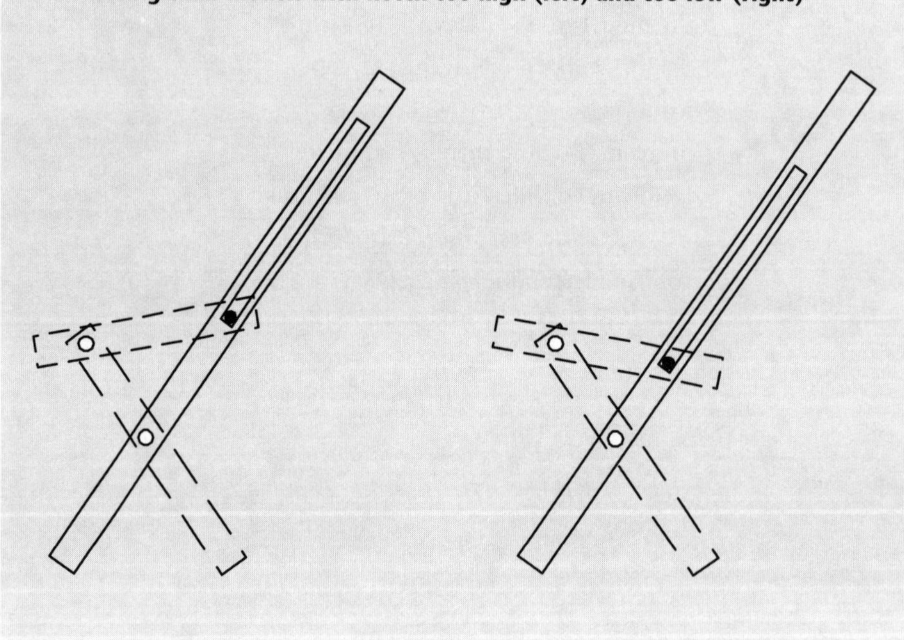

What should the links be made of?

A link is simply a rigid piece, such as one of the rectangles on the folding chair model in Figure 1-33. Links should be made of something that is fairly stiff and strong, not too difficult to cut, and very inexpensive or (better yet) free. Cardboard is the only material we have found that meets all of these criteria.

Figures 1-35 A, B, C, D, and E show four of the most common types of cardboard, all of which are useful for modeling mechanisms. A is "card stock," used to make index cards, file folders, and other items that don't need to be very stiff. B and C are made of heavier grades of flat cardboard, and are used to package food and other small items. D is a thin grade of corrugated, often found in pizza boxes, shoe boxes, and the like. E is a common grade of corrugated cardboard, found in medium-sized cartons. All five types of cardboard can easily be cut with a scissors or punched with a standard hole punch.

As you begin to experiment with these materials, you will notice some major differences among them. A strip of card stock, for example, cannot be pushed too hard from one end, or it will buckle. On the other hand, the thicker corrugated grades are harder to join, because the pivots need to be longer and the rougher surfaces do not slide as easily.

An interesting feature of corrugated cardboard (Figure 1-35E) is that its properties depend on the way it is oriented. Cut a small strip of cardboard as shown in Figure 1-35F. If you push both ends parallel to the corrugations, you will find it much stronger than if you push at right angles to them as in Figure 1-35G.

1-35: **Five types of cardboard**

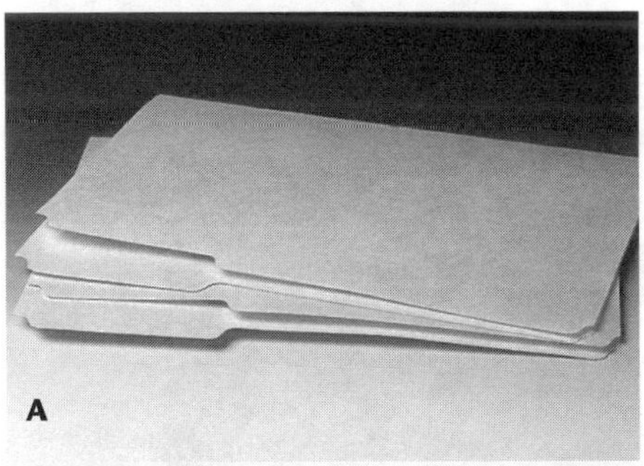

How can you make a pivot, which allows one link to rotate with respect to another?

The open circles on the folding chair model in Figure 1-33 are pivots. The brass paper fastener (Figure 1-36) is very useful for making a pivot between two links or for attaching a link or guide to a cardboard base.

Begin by using a hole punch to make holes in the two links to be joined. When you open the two tails and flatten them, the heads and tails should prevent the fasteners from slipping out. Their major drawback is that their shafts are flat rather than round, and the links may not be able to move freely around them.

Paper fasteners come in several sizes, ranging from 1/2 to 1 1/2 inches long. They should be large enough so the heads don't slip through the holes, but not so large that the flattened tails interfere with one another when the fasteners are close together. Figure 1-37 shows a model of a retaining ring pliers made from paper fasteners and thin corrugated cardboard. The actual tool is shown in Figure 1-38.

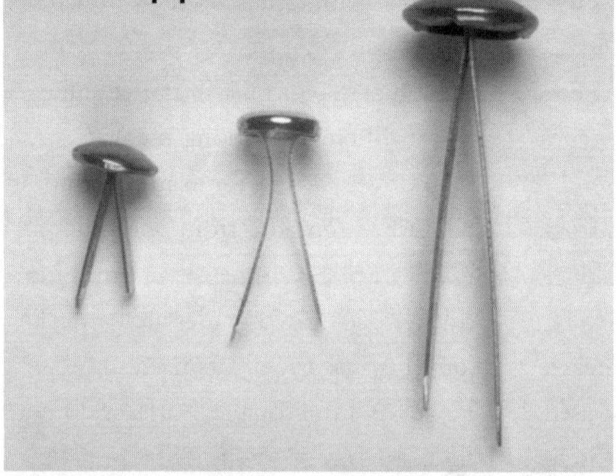

1-36: **Brass paper fasteners**

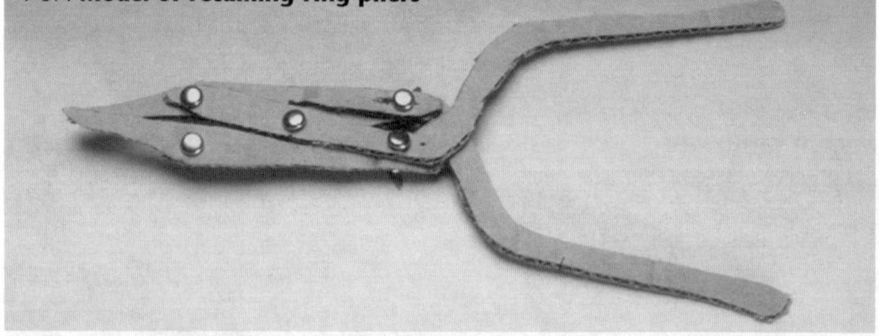

1-37: **Model of retaining ring pliers**

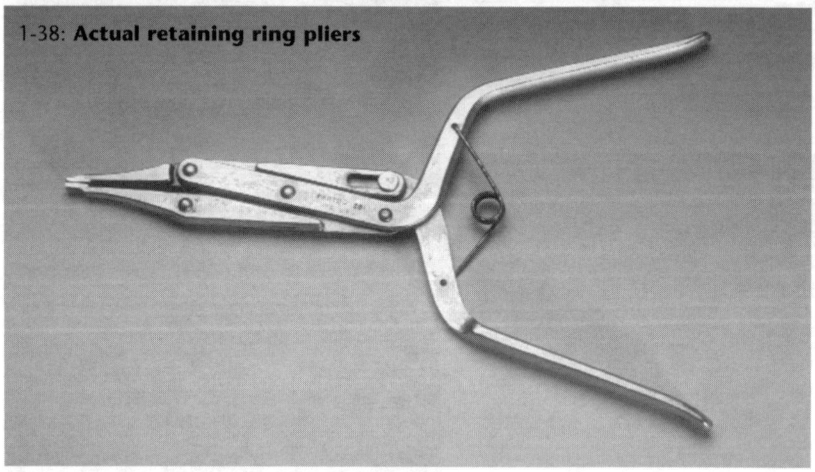

1-38: **Actual retaining ring pliers**

Another technology we have found for making pivots is the eyelet—the same kind that is used in eyeletting fabric. These are sold in craft stores, along with a special tool for crimping them, which costs about $10. (See Figure 1-39.) As with paper fasteners, it is necessary to punch holes first, slightly larger than the eyelets; otherwise, the links will not be able to rotate. Unlike paper fasteners, eyelets are permanent—they cannot be removed once they have been installed. Also they do not work with all grades of cardboard—card stock is the best. However, unlike paper fasteners, their shafts are round so the links can move freely, and they look much better than paper fasteners. Some teachers have been very successful with eyelets, but they are more difficult to use than paper fasteners.

1-39: **Eyeletting tool and eyelets**

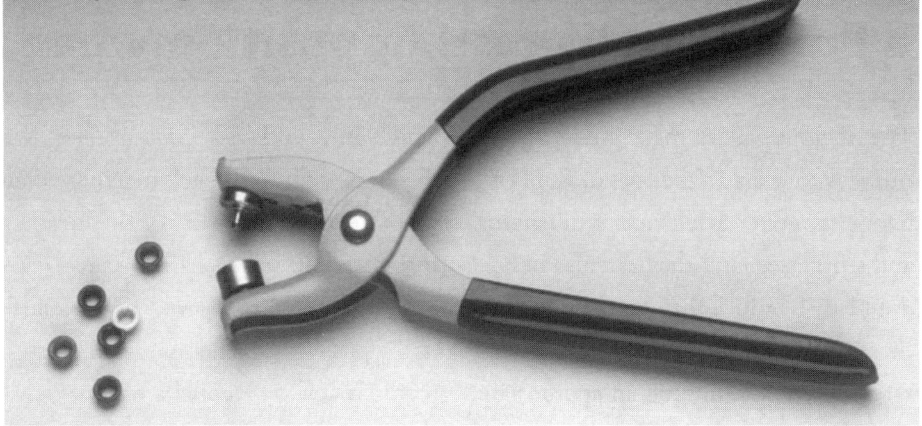

How can you make a sliding joint, which permits a link to travel along a straight line?

The black circular pin-in-slot on the folding chair model in Figure 1-33 is an example of a sliding joint. There are two easy ways to make sliding joints. One involves attaching a guide to the base on either side of a link to force it to follow a straight line, as shown in Figure 1-40. The circles are paper fasteners, which hold the guide in place. A little space should be left between the sliding link and the guide, so that it can move freely. The other method uses a slot in the base, with paper fasteners holding the front and back of the sliding link together. (See figure 1-41.) The extra piece of cardboard on the back allows the slider to move much more easily, and makes the width of the slot much less critical.

1-40: **Sliding joing using a guide**

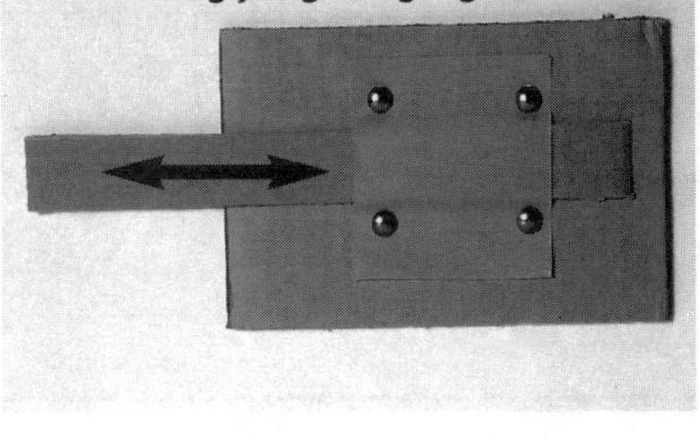

1-41: **Sliding joint using a slot and paper fasteners (back and front views)**

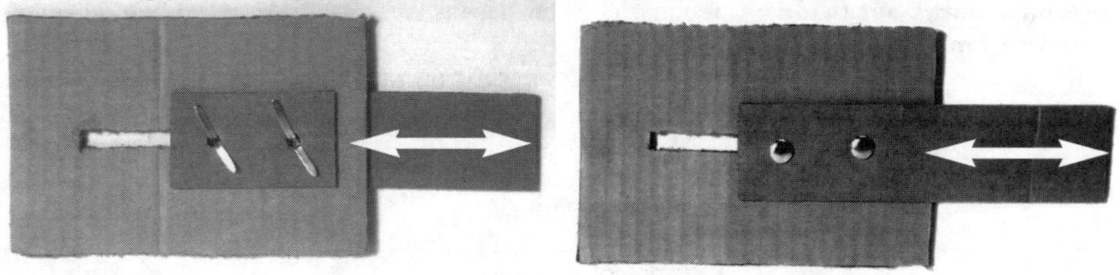

What can you use to make a return spring, which restores a linkage to its resting position when it is released?

The obvious way to make a *return spring* is to use a rubber band. Rubber bands come in a wide variety of lengths and stiffnesses, and they can also be doubled to reduce the length and increase the stiffness; therefore there is rarely a problem finding an appropriate one. One little problem is that rubber bands only work as *tension springs:* the ends try to pull back together when you extend them. Many of the metal springs found in mechanisms are actually *compression springs* whose ends try to push apart when you compress them. When you model a return spring, you may need to keep this problem in mind, and mount the rubber band in a different place than the corresponding compression spring. A bigger challenge is finding good places to attach a rubber band. One effective method is to loop it around two paper fasteners.

Model Troubleshooting Guide

A model rarely works perfectly the first time. In a way that's good, because most of what you learn from model-making happens during the process of troubleshooting and redesign. This troubleshooting guide covers some of the problems commonly encountered in modeling mechanisms.

Table 1-2

TROUBLESHOOTING GUIDE

Symptom	Possible Cause	Remedy
Model can't reach extreme positions	Links are not made to scale	Trace parts directly on cardboard; begin with actual size model
	Joints are not positioned properly	Use chalk or marker to transfer joint positions to model
No return action when released	Return spring missing	Add rubber band
	Return spring too weak	Use stronger rubber band or double it
Inputs or outputs do not move in straight line	Sliding joints are missing	Use guides or slots to constrain motion
Model works but is too hard to move	Too much friction in joints	Make larger holes (piviots) or wider slots or guide openings (sliders)
	Paper fasteners are colliding with one another	Use smaller fasteners, or turn them around
Links buckle or bend when model is operated	Links are not strong enough	Use heavier cardboard for links
	Too much friction	See above
	Parts do not fit properly	Make parts to scale

Mystery Mechanisms: Design Puzzles

1-42: **Design Puzzle #1**

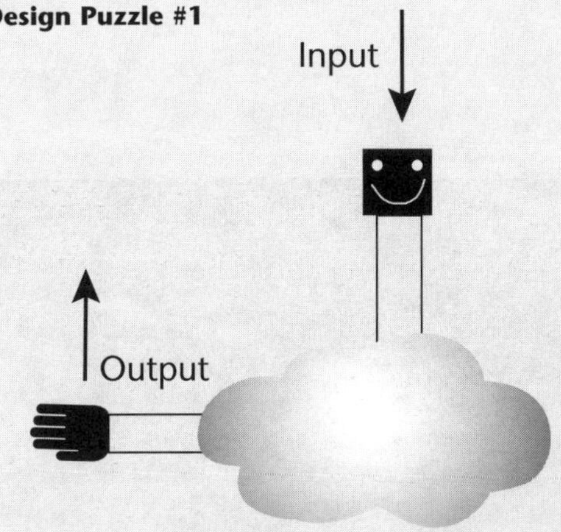

1-43: **Design Puzzle #2**

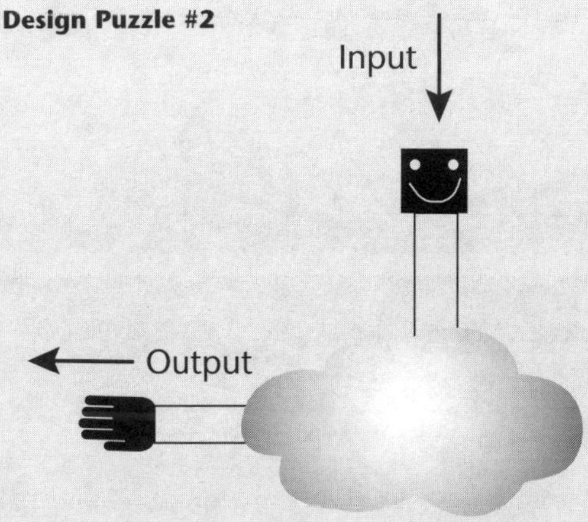

1-44: **Design Puzzle #3**

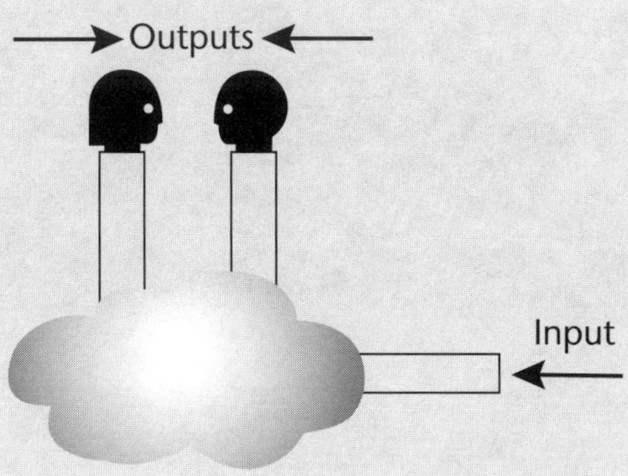

All of these mechanisms can be made using the modeling techniques just described. As you work on the puzzles, look through your mechanism collection for devices that involve similar problems. Each problem is presented by showing the inputs and output(s) only. It is your job to design and test the mechanism that connects the input to the output. The solutions to all three problems are given in the next chapter.

Design Puzzle #1:
The arms flap up when the head goes down!

Imagine a doll with a movable head and two movable arms. When you press down on the head, the arms move up. To solve this problem, begin with one side only. The input and output are shown in Figure 1-42.

Design Puzzle #2:
The arms shoot out when the head goes down!

This problem is similar to the first, except that instead of moving up, the arms shoot out when the head is depressed. Again, start with one side only. The input and output are shown in Figure 1-43.

Design Puzzle #3:
The "kissing couple" problem

Your task here is to make a little toy that has an input on the side, and two outputs on the top. When the input is pushed in, the two outputs come together. We call it "the kissing couple" problem because some children have attached heads on the two output links. They appear to be kissing when the input is operated. The input and output are shown in Figure 1-44.

Casting a Bright Light on Electric Circuits

The Switch Connection

Here is another mystery. Figure 1-45 shows an ordinary flashlight. When you slide the switch forward, as in Figure 1-46, the flashlight comes on. What is going on inside the flashlight when you slide the switch forward? Why does the bulb light when the switch is in the forward, but not the rear, position?

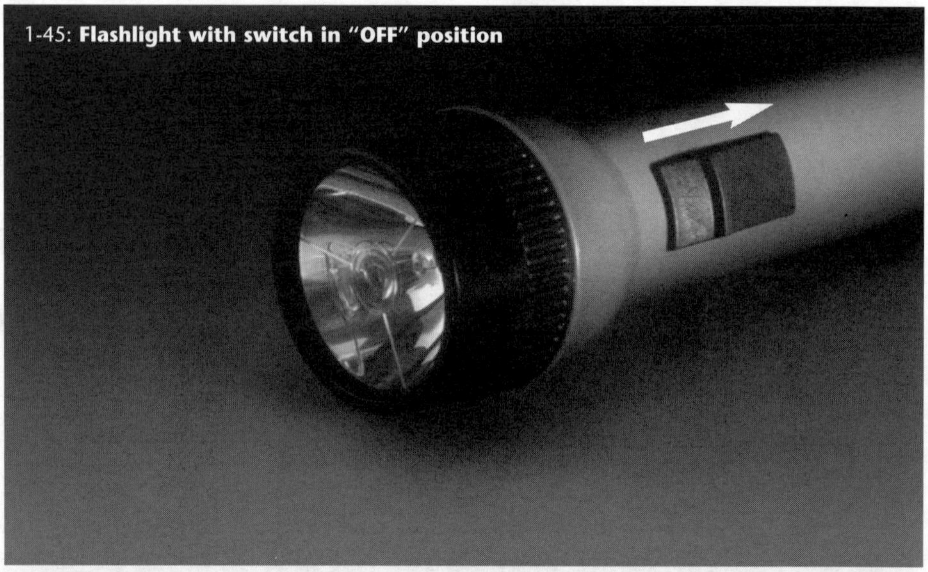

1-45: **Flashlight with switch in "OFF" position**

1-46: **Same flashlight, turned on**

1-47: **Inside the flashlight, showing switch in "OFF " position**

1-48: **Switch "ON" showing clip forced towards center**

1-49: **Front cover, showing cup that holds the bulb**

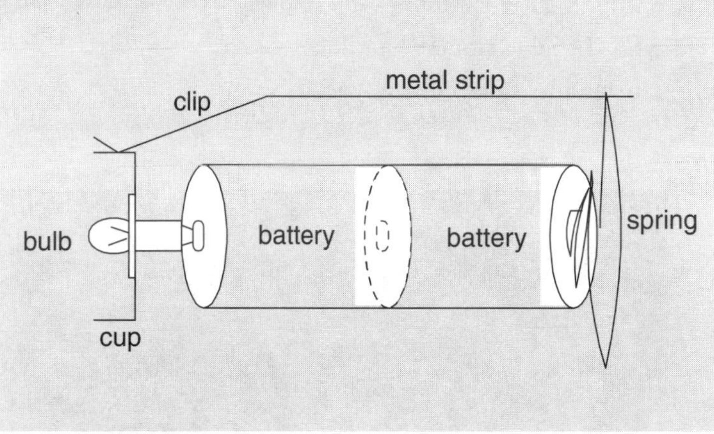

1-50: **Circuit of flashlight, in "ON" position**

If you remove the cover and the batteries as shown in Figure 1-47, you can see exactly what happens. There is a little copper clip (indicated by arrow) attached by a metal strip to the spring that sits behind the batteries. When the switch is moved forward, it pushes the front of the little clip towards the center (Figure 1-48). This action forces the clip in contact with the metal cup that holds the bulb (Figure 1-49). Now there is a complete *circuit* consisting of the two batteries, bulb, clip, metal strip, and spring (Figure 1-50).

Meanwhile, the tip of the battery is in contact with the end of the bulb, which is its other *terminal.*

This description shows how the switch works mechanically, but it doesn't explain anything at all about how and why electricity makes the bulb light up. We will postpone this discussion to the next chapter. The focus here is on the one mechanical component of the flashlight: the *switch.* Switches of one kind or another control nearly all electrical circuits, and it is worth looking at them closely.

Switch Scavenger Hunt

Most switches have two positions, ON and OFF. The switch that controls the flashlight in Figure 1-45 is an ON/OFF switch that slides back and forth between its two positions. You can leave it permanently in either position; unlike some switches, it does not spring back. There are a variety of common designs of ON/OFF switches, of which the flashlight's "slide switch" is only one kind. Here is a list of some popular types. Looking around your house, can you find some of each type?

- **Slide switch** (Figure 1-45);
- **Rocker (or toggle) switch:** swings in a slight arc, like a rocking chair—most wall switches are of this type (Figure 1-51);
- **Rotary switch:** uses a knob or handle to turn in part or all of a complete circle—often used in conjunction with a dimmer or volume control (see below), or for a situation in which there is more than one ON position (Figure 1-52);
- **Pushbutton switch:** like a slide switch, it travels in a straight line, but in-and-out, rather than side-to-side (Figure 1-53);
- **"Radio" buttons:** another variation of the pushbutton, made famous in car radios, where only one of several buttons can be ON at any one time (Figure 1-54).

1-51: **Rocker (toggle) switches**

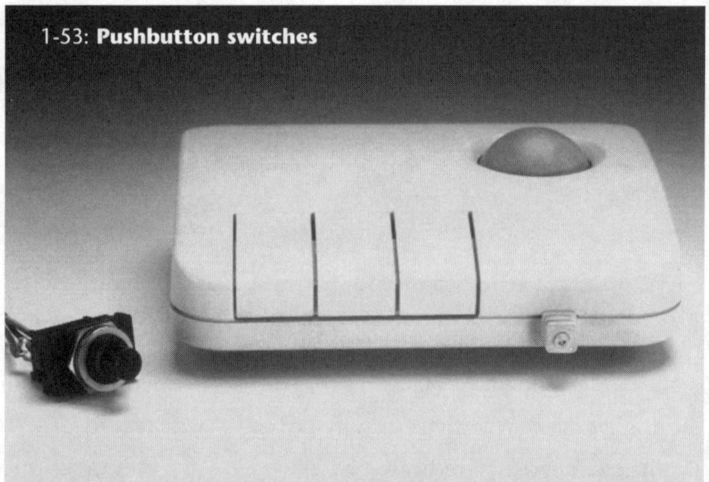
1-53: **Pushbutton switches**

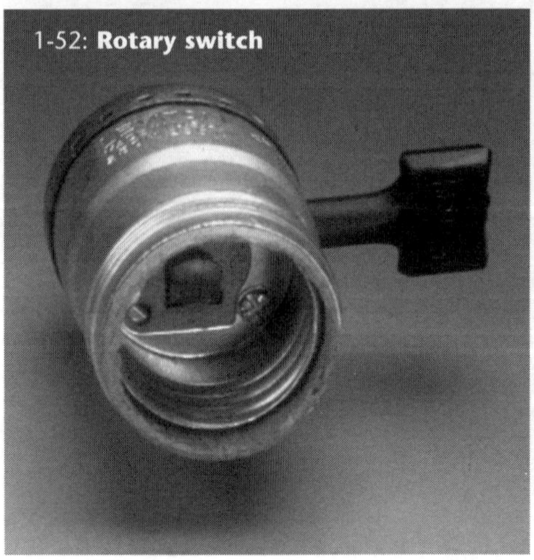
1-52: **Rotary switch**

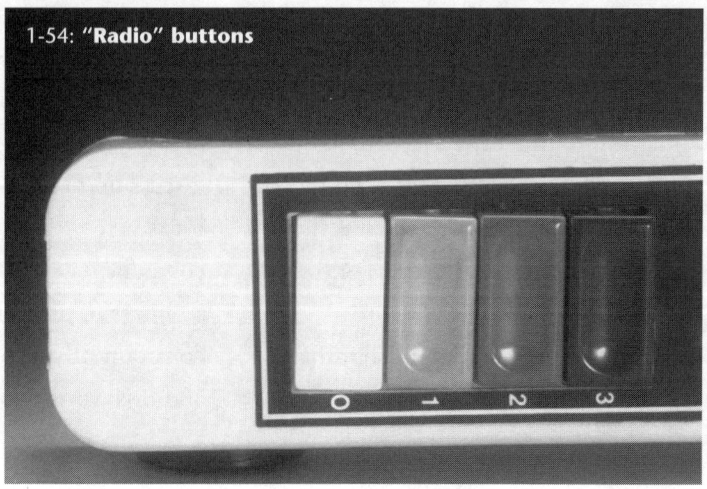
1-54: **"Radio" buttons**

Besides the categories suggested above, what other ways might there be of classifying switches? One question to ask is: What does it control? Each of your switches controls a device that uses electrical energy to produce some kind of output, such as light, heat, sound, or motion. Another way to classify begins with the observation that each switch is also a mechanism. Some of them use levers, while others have sliders; some have return springs, while others don't.

From Switches to Controls

One important category consists of switches that are concealed. Here, the user operates the switch without intending to. Nearly always, these are momentary switches, which are returned by a spring to the OFF position when the device is not in use. For example, when you insert a pencil in an automatic pencil sharpener, the pencil activates a concealed switch. This switch closes a circuit that turns the motor on. The switch returns to its OFF position when the pencil is removed. (See Figure 1-55.) Other examples of concealed switches are:

- the switch under the brake pedal of a car, which operates the brake lights;
- the switch behind a car door that turns on the dome light;
- the switch that turns on the light in the refrigerator; and
- the switch that turns on a telephone when the receiver is lifted.

Try to locate each of these concealed switches and operate them by hand.

Not all switches select between only two positions. Most fans have at least two ON positions, labeled HIGH and LOW. This is also the case with most hair dryers, clothes dryers, blenders, mixers, and some other appliances that use motors. Some of these devices are controlled by two separate toggle switches, one for ON/OFF. and the other for HIGH/LOW, as in many hair dryers. Other devices use radio buttons for the same purpose (as an example, see the floor fan switches in Figure 1-54). In other cases, a single rotary switch is used to select the setting. (See Figure 1-56.)

1-55: **Concealed switch inside automatic pencil sharpener**

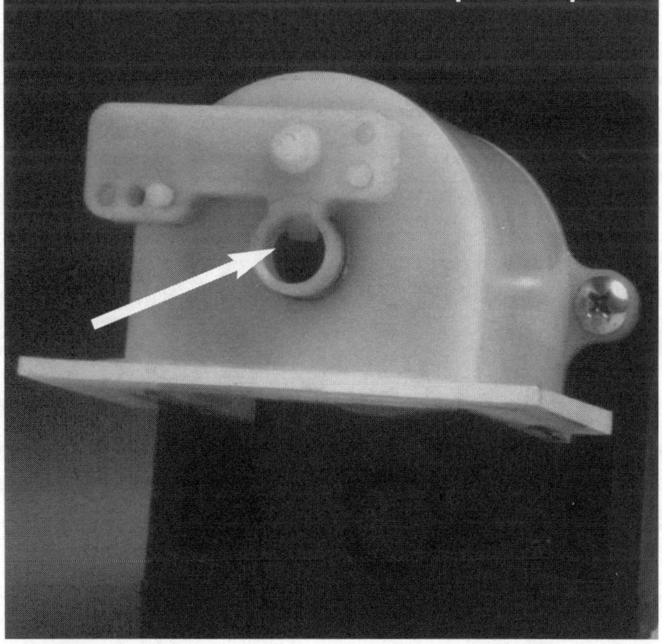

Sometimes, a knob or slide mechanism can vary something continuously, rather than select from among a few possibilities. These devices are like switches in that they can turn something on or off, but they are more like faucets in that they can be adjusted to any position between fully on and fully off. (See Figure 1-57.) Anything that can be varied continuously is known as an *analog device,* in contrast with a *digital device,* which has a finite number of discrete settings. A tuning knob on a radio is analog, while the programmable push buttons—referred to earlier as radio buttons—are digital.

A word that includes both the digital "switch" and the analog "faucet" is a *control.* In this section, we will look specifically at electrical controls—those that adjust the flow of electricity, while in the next section, we will expand the discussion to include other kinds of controls. In the next chapter, "Concepts," we will define and explain the concept of control more precisely.

The preceding discussion suggests some new categories for the scavenger hunt. How many different examples can you find of each of the following?

- multi-position controls that select from more than one possible ON position
- analog controls, which can be adjusted anywhere from fully ON to fully OFF
- hidden controls, which are operated without the user necessarily knowing about them.

What kinds of mechanisms—rotary, toggle, slide, pushbutton, or momentary—are most frequently used for each kind of control?

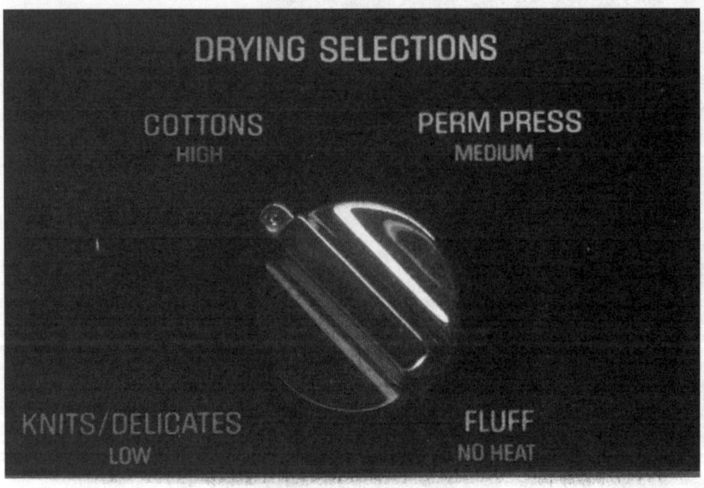

1-56: **Rotary multi-position switch**

1-57: **Continuously variable controls**

Electric Circuit Challenges

As with mechanisms, an excellent way to learn about circuits is to model them. We conclude this chapter with a few more challenges, this time involving circuit models. We are presenting these challenges in increasing order of difficulty. Don't worry about what the models looks like. The important thing is that they should work like the real thing.

You will need the following materials: at least one battery, at least one bulb, and some wire, preferably insulated. (See Figure 1-58.) If the wire is insulated, you will need a wire cutter/stripper to remove the insulation at the ends, as shown in Figure 1-59. These components are ordinarily supplied with science units on "Batteries and Bulbs." It is helpful, but not absolutely essential, to have bulb holders and battery holders, to make it easier to connect wires to these components. If bulb and battery holders are not available, just use a little tape to secure the stripped end of a piece of wire to the contacts, as shown in Figure 1-60.

You will also need a switch, but if you don't have it, you can always make one. The switch can consist simply of two wires that are touched together for one position and separated for the other. Better yet, you can make a permanent switch using anything made of fairly flexible metal, such as a paper clip, hairpin, or piece of aluminum foil, as shown in Figure 1-61.

1-58: **Battery, bulb, and wire**

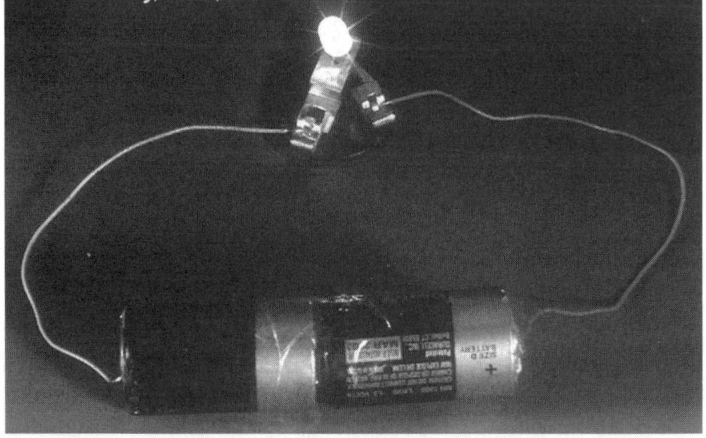

1-59: **Using a wire stripper to remove insulation from the end of a piece of wire**

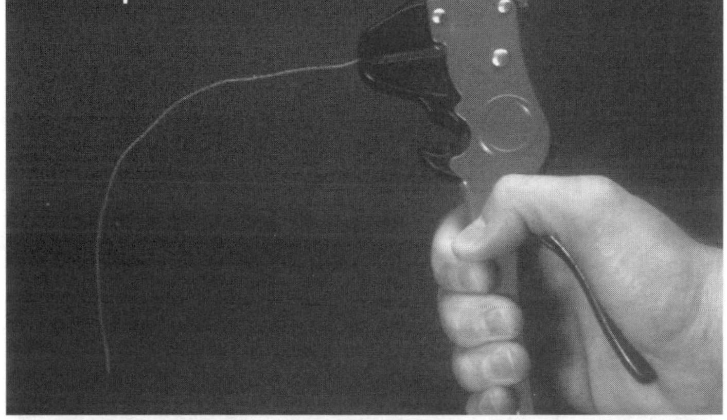

1-60: **Connecting to batteries and bulbs without bulb holder**

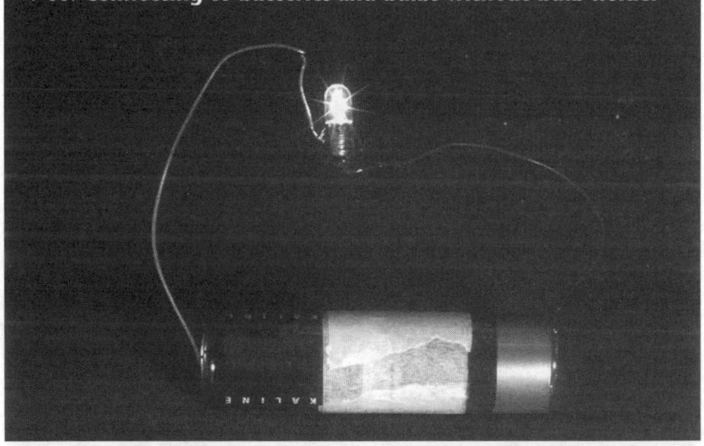

1-61: **Home-made switch**

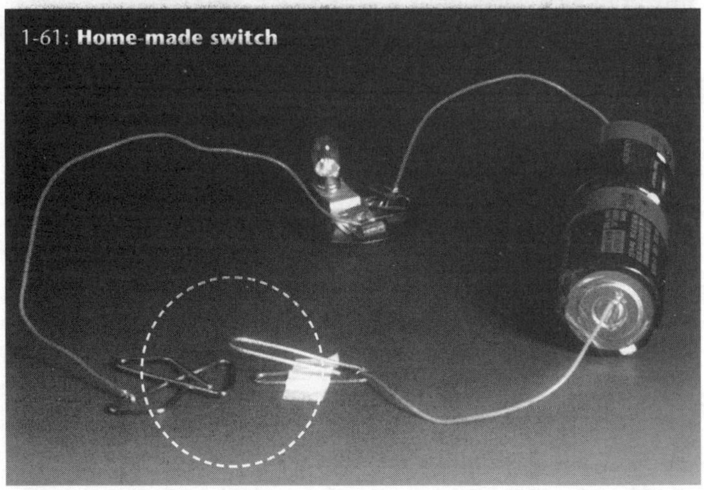

The Flashlight

Begin by making a working model of a flashlight, whose switch controls the lighting of the bulb. The bulb should light when the switch is in one position, but not when it is in the other position. If you have trouble getting it to work, the next chapter will help, but first you should try it for yourself!

The remaining circuit challenges are more difficult, and you should not attempt them until you have solved the flashlight problem. We will describe three situations from everyday life where two switches need to control a single bulb. In each case, the logical relationships between the switches are different. The questions each time are:

- Can you think of other examples where the same logic applies?
- Can you model this situation using batteries and bulbs?

The Lamp Situation

Suppose a lamp with a switch on it is plugged into a power strip, which has its own switch. Both switches have to be in the ON position for the light to come on. The same description applies when a lamp with its own switch is plugged into a wall socket, which is controlled by a wall switch elsewhere in the room. The lamp will not come on if only one or none of the switches is activated, but only if both are.
The Lamp Situation is summarized in Table 1-3.

Can you think of other systems that have the same logic as the Lamp Situation? These would be devices that are controlled by two switches, both of which have to be ON for the device to come on. You can also make a model of this situation. If you were successful in modeling the flashlight, you might want to try modifying your model to incorporate an additional switch. Can you arrange the two switches so they obey the logic shown in Table 1-3? The solutions to this problem and the next two are in the next chapter, but first try to do them yourself!

Table 1-3

THE LOGIC OF THE LAMP SITUATION

Switch on Power Strip	Switch on Lamp	Lamp
OFF	OFF	OFF
OFF	ON	OFF
ON	OFF	OFF
ON	ON	ON

The Yard Light Situation

A floodlight is set up to illuminate the back yard of a house. There are two switches that operate the yard light, one inside the house and the other in the yard. Either switch (or both at the same time) will make the yard light come ON. This situation is summarized in Table 1-4.

Notice the differences between Table 1-3 and 1-4. Table 1-3 has only one combination that turns the lamp on, while Table 1-4 has three. In the Lamp Situation, both switches need to be ON to make the lamp come on, while in the Yard Light Situation, either one, or both, will do the trick. Another example of this situation is a two-door car with a dome light. In this case the switches are concealed. Opening either door, or both, will make the dome light come on. Can you think of other examples?

Table 1-4
THE YARD LIGHT SITUATION

Indoor Switch	Outdoor Switch	Yard Light
OFF	OFF	OFF
OFF	ON	ON
ON	OFF	ON
ON	ON	ON

The Stair Light Situation

The third configuration, the Stair Light Situation, is also fairly common, but considerably more difficult to model. If your house has more than one floor, it is likely that you have a ceiling light at the top of a stairway. Two switches usually control this type of light—one at the top of the stairs and the other at the foot. Either switch can turn the light ON. So far, this sounds exactly like the Yard Light Situation, but there is a difference. Either switch can also turn the light OFF, which is not the case in the Yard Light Situation.

To see how the stair light works, imagine that you are at the top of the stairs. Both switches are in the DOWN position, and the light is OFF. To see your way down the stairs, you push the top switch UP, turning the light ON. When you reach the bottom, there is no longer any need for the light, so you push the switch at the bottom UP, turning the light OFF. This situation is similar to the Yard Light, in that either switch, alone, will turn the light ON, but different in that activating both of them will turn it OFF again. The Stair Light Situation is summarized in Table 1-5. You can compare Tables 1-4 and 1-5 directly. Note that the Stair Light Situation is identical to the Yard Light situation, except for the case where both switches are ON.

Table 1-5
THE STAIR LIGHT SITUATION

Upstairs Switch	Downstairs Switch	Stair Light
OFF	OFF	OFF
OFF	ON	ON
ON	OFF	ON
ON	ON	OFF

Chapter 2

CONCEPTS

Children are naturally curious about and fascinated by mechanisms. They love to take things apart to reveal the secrets hidden inside. That's one reason the study of mechanisms is an engaging and inexpensive way to introduce several of the big ideas of science and technology to young children. This chapter describes the concepts—those big ideas—that children are exploring when they investigate mechanisms. You'll revisit the challenges and puzzles you worked with in Chapter 1 in the context of these concepts and children's learning. In Chapter 3, you'll find activities that let children use their hands and their minds to explore these ideas directly. And in Chapter 4, you'll see how some teachers applied these ideas in their classrooms through their work with mechanisms.

Mechanisms Are Systems

A *system* is a collection of interconnected parts, functioning together in a way that makes the whole greater than the sum of its parts. A *mechanism* is a particular kind of system—one that converts one type of motion to another. A bicycle has one mechanism that converts pedaling to forward motion, another mechanism for operating the brakes, and yet others for changing gears. The human body is itself a system that abounds in smaller systems—including mechanisms that use muscles to transmit motion to limbs and other body parts.

By looking at mechanisms as systems, even young children can begin to build their understanding of how the world works by handling questions like these:

- What is the purpose of this system as a whole? What does it do?
- What are its parts?
- How are the parts connected to one another?
- How does each part function in relation to the whole?

Every mechanism can be understood in terms of these three basic elements:

- the *input*—the motion that causes the system to operate;
- the *output*—the motion that is the result or effect of the input;
- the process that changes the input into the output.

Look at a glue stick, for example. It is a simple mechanical system that takes one kind of motion and transforms it into another. Remove the cover from a glue stick and examine the tube. To use the glue, you turn a knob at one end and the glue comes straight out the other end. (A lipstick tube works the same way.) Inside the tube is a mechanism that takes the motion you supply by turning the knob—the input—and transforms it into the motion that makes the glue come out—the output. The input and output motions are different. The input motion travels around in a circle at one end of the glue or lipstick case. The output motion travels in a straight line and moves most of the length of the stick.

2-1: **A glue stick as a mechanism**

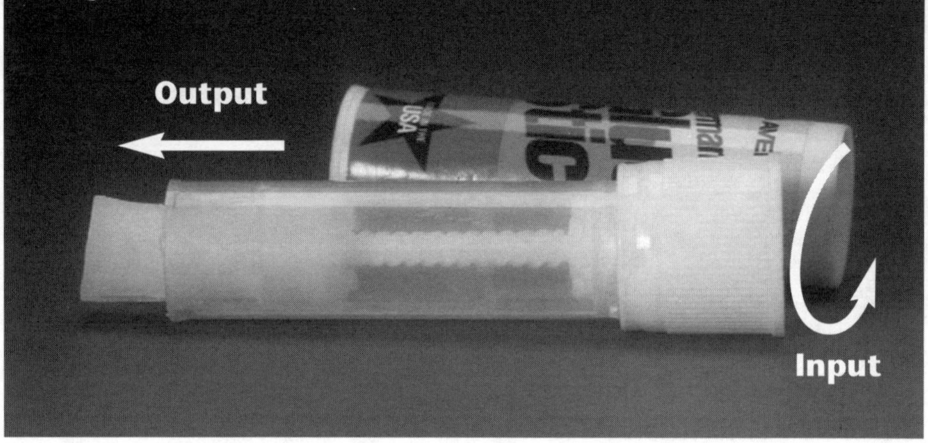

Making Things Happen

Unless a system is broken, it has a process that converts the input into the output. Another way of saying this is that there is a cause-and-effect relationship between the input and the output. The idea that one thing causes another is the basis for scientific prediction: if the cause is present, you can predict that the effect will occur.

As Piaget demonstrated, even young children have very definite ideas about cause and effect, but many of these ideas are at variance with the conclusions of science. When he asked five- through eight-year-old children what makes clouds move, their answers included "by themselves," "God," "the sun," "the moon," "rain," "night," "the air from the trees," etc.

In science education, cause-and-effect is often introduced by developing the formal method of a controlled experiment. For example, if some classroom plants thrive better than others, it might make sense to do a controlled experiment to determine precisely why. The controlled experiment is a method of establishing cause-and-effect when the cause is not immediately obvious. However, children below the upper elementary grades may not be developmentally ready to handle the concepts of a variable, a controlled variable, an experimental variable, and so on.

Simple mechanisms offer more direct opportunities for learning about cause-and-effect. Just by playing with a mechanism, even very young children can usually figure out what causes what. By choosing mechanisms of increasing complexity, teachers can introduce children to more complicated examples of cause-and-effect sequences.

Taking It Apart, Piece by Piece

What does it mean to say, "I understand how this mechanism works"? Let's consider the example of a glue stick again. Some brands of glue stick have a case that is semi-transparent, as shown in Figure 2-2. As you turn the knob at the right, you can see that it turns a screw inside the tube that runs the length of the stick. As the screw turns, a kind of nut rides left or right on it, depending on the direction of turning. The glue sits in a platform attached to the nut, so that turning the knob and screw forces the platform to move left or right.

2-2: **Turning the knob at the bottom of the glue stick turns the screw inside the tube.**

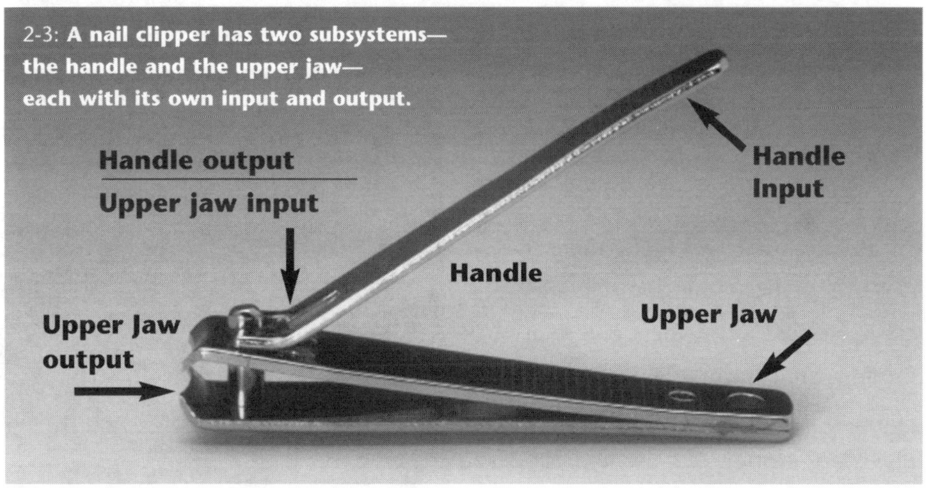

2-3: **A nail clipper has two subsystems—the handle and the upper jaw—each with its own input and output.**

This explanation of a glue stick consists of a step-by-step description of what all the parts do, how they transmit motion from one to the next, changing the input into the output.

A nail clipper is somewhat more complicated than a glue stick. It has about the same number of parts as a glue stick, but the parts are not as tightly connected and their relationships are not quite as clear.

In analyzing a nail clipper, it is useful to divide the device into two sub-systems or modules—the handle and the upper jaw—each with its own input and output. (See Figure 2-3.) The output of the handle is the input to the upper jaw. In other words, the two parts are in series: the tail of one is attached to the head of the next. In this example, a more complicated mechanism can be analyzed by decomposing it into a sequence of simpler mechanisms.

Dividing a complex system into simpler ones is a very important technique in dealing with systems, and for problem solving in general. The same idea is behind modular home construction, stereo component systems, software "add-ons," and solving math problems by the "divide-and-conquer" strategy. It is also a basic strategy in design: solve the problem one part at a time, and then combine the parts.

Science and Math Concepts in Mechanisms

Types of Motion

The purpose of a mechanism is to change some aspect of the motion at the input so that it is somehow different at the output. There are two basic types of motion: *linear motion* and *rotary motion.* When an object moves linearly, it moves along a straight line. Rotary motion, on the other hand, follows a circular path, which may or may not complete a full circle. Linear motion in one direction is called *translation.* Clockwise or counterclockwise rotary motion is known as *rotation.*

Besides these two types, there are two other types of motion that are closely related. Pure linear and rotary motions are continuous—they keep going in the same direction until they stop. Motion can also go back-and-forth. Back-and-forth motion in a straight line is called *reciprocating motion,* while rotary back-and-forth motion is known as *oscillating motion.* A summary of all four types, with examples, can be found in Table 2-1.

Some mechanisms are designed to convert one of these four types of motion to another. For example, the screw mechanism in a lipstick container or glue stick converts the rotation of the knob to the translation of the lipstick or glue. Something similar happens in a door lock, which converts

Table 2-1

TYPES OF MOTION

Type	Description	Diagram	Examples
Translation	Linear continuous	→	zipper, slide bolt, push button
Rotation	Rotary continuous		Steering wheel, fan blades, faucet handle, can opener
Reciprocating	Linear back-and-forth	↔	sewing machine needle, bicycle pump, piston
Oscillating	Rotary intermittent		windshield wipers, oscillating fan, lawn sprinkler

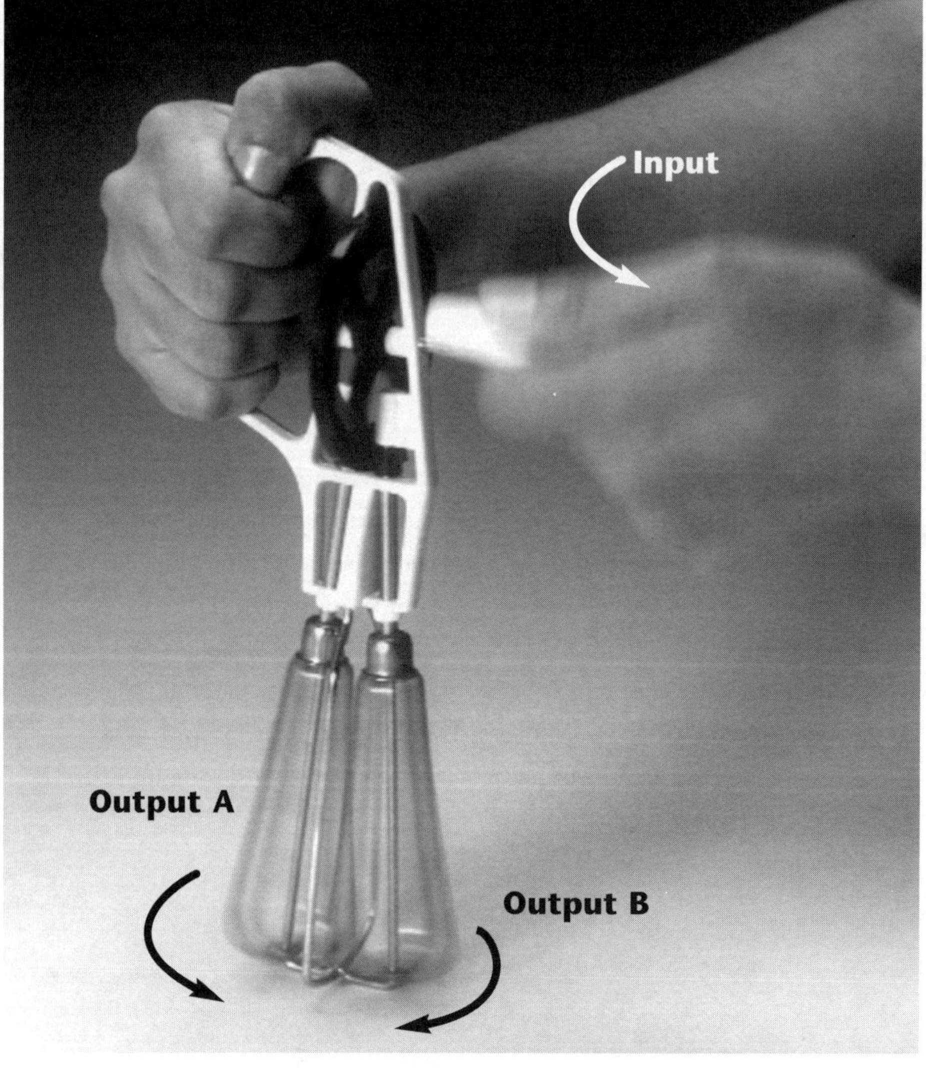

2-3: **In the eggbeater, all the motions are rotations, but they go in different directions and at different speeds.**

the rotary motion of a key to the linear motion—the translation—of the bolt. A windshield wiper linkage converts the rotary motion of a motor to the oscillating motion of the wiper blades.

There are also many mechanisms that do not change the type of motion, but rather change its speed, direction and/or location. In the eggbeater shown in Figure 2-3, all the motions are rotations: two at the output and one at the input. The eggbeater takes one rotation and produces two rotations in opposite directions. Notice that the outputs are different from the input in several ways:

- There are two outputs and only one input;
- The two outputs go in opposite directions;
- The input rotation is vertical, while the output rotations are horizontal;
- The outputs rotate faster than the input.

Mechanisms are often designed to magnify the force available for a job. The nail clipper of Figure 2-2, for example, transforms a relatively small amount of effort at the handle into the much greater force needed to cut the nail. This transformation depends on the use of levers, which we shall explore in detail shortly.

"Ins and Outs of Inputs and Outputs" in Chapter 3 is an activity designed to explore types of motion.

Links and Joints

Most of the common mechanisms we will consider are composed of two kinds of components: links and joints. A link is a rigid bar, frame, or plate that can move only as a unit. Consider, for example, the folding chair shown in profile in Figure 2-4. This is the same device shown in Figures 1-33 and 1-34. It consists of three links: the back, the seat and the diagonal brace.

The chair could never fold and open if these links were simply glued together. Instead they are connected by joints, which allow rotary or linear motion. In the diagram, you can see the three joints, represented by circles. Two of the joints—those connecting the seat and brace, and the brace and back—are pin joints, which only allow rotation. Some other examples of pin joints are the human elbow, a door hinge, and a toilet-paper-roll holder.

The third joint of the folding chair—the one connecting the back and seat—is a roll-slide joint. Notice how the pin-in-slot arrangement allows both translation and rotation. If you compare the positions of the seat in the open and closed positions, you can see how the seat has to both rotate counter-clockwise and slide up in order to fold up. You can find more roll-slide joints in umbrellas, scissor jacks, foot-pedal pumps, ironing boards, and drafting tables. A folding umbrella has a slider that rides along the central shaft. Attached to the slider are links that are vertical in the folded position, and gradually become horizontal as the umbrella unfolds. Each of these links is connected to the central shaft by a roll-slide joint. (See Figure 2-5.)

2-4: **Folding chair in open and closed positions**

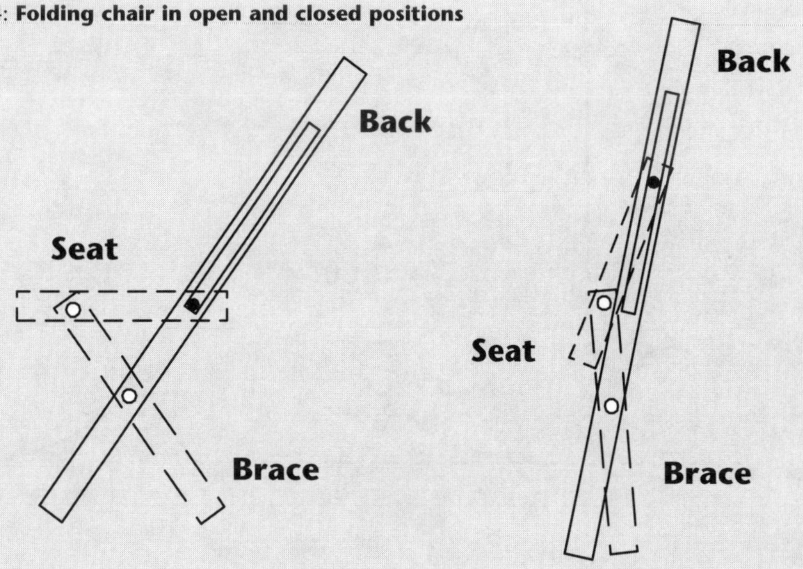

Some joints are not as permanent as door hinges or the pin joints in folding chairs, which stay together because of the way they are made. They are temporary joints that work only as long as there is a force pushing the links together. An example occurs when you use a hammer to remove a nail (Figure 2-6). The hammer rotates against the fixed surface, but only because it is held there by a force. The force is gravity if it is resting on the floor (Figure 2-6A) or the force of your hand if the hammer is against the ceiling (Figure 2-6B).

The "Make a Model of a Mechanism" activity in Chapter 3 is designed to explore links and joints.

2-5: **Detail of a folding umbrella in closed and open positions**

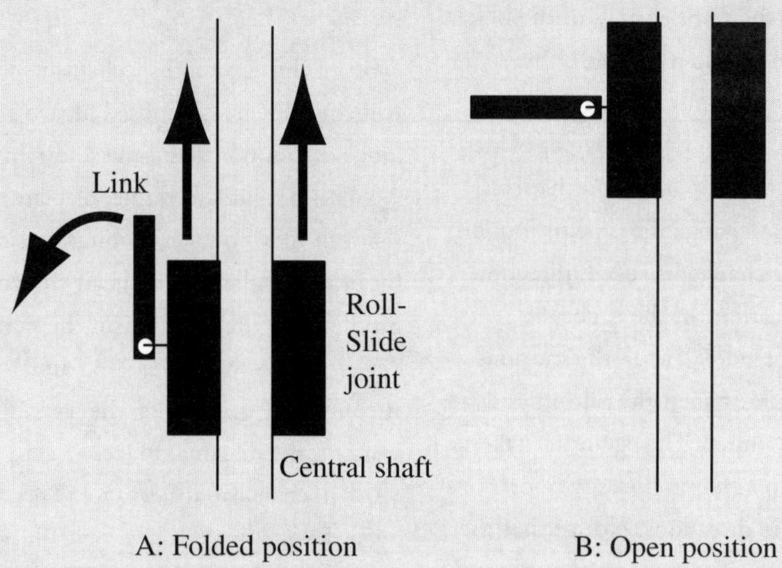

2-6: **Temporary joints formed by a hammer held by force against a fixed surface**

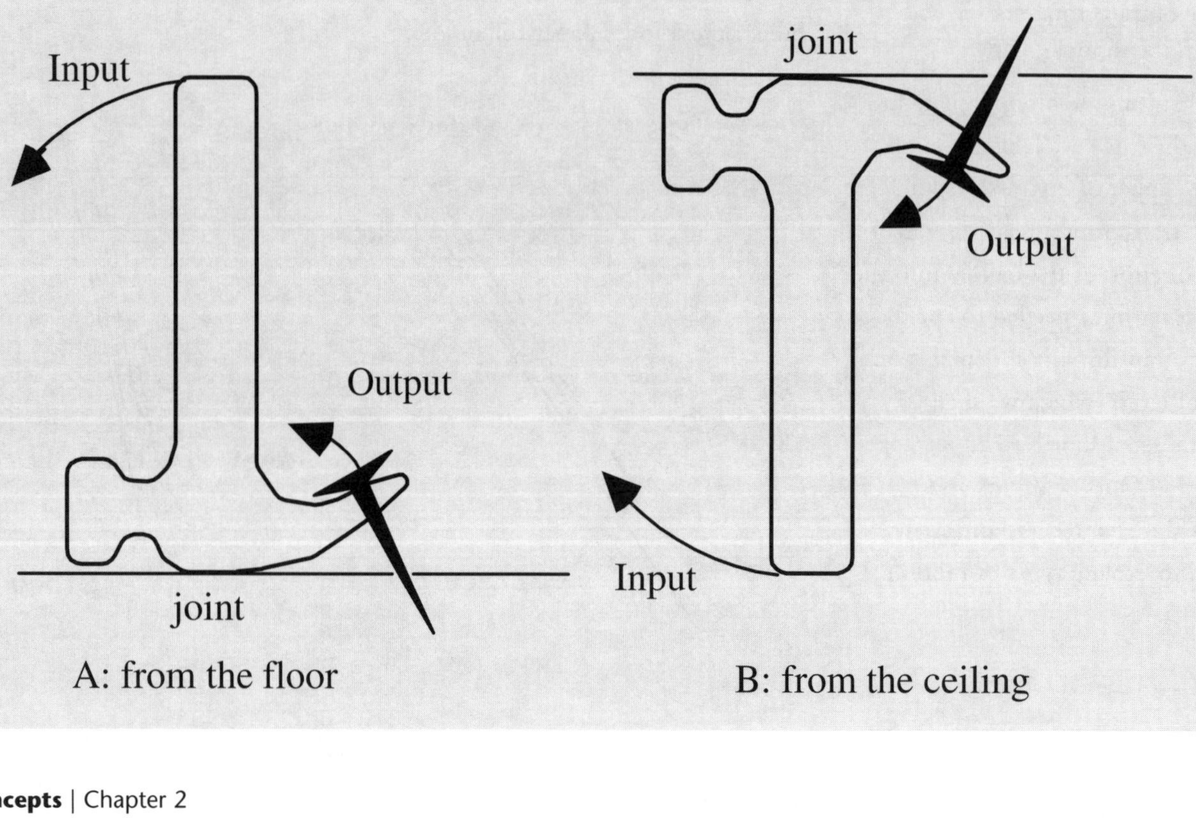

How Levers Work

Suppose you have to lift a heavy desk momentarily to get something out from under it. One way is to use a long wooden board to pry it up. Near the desk end, rest the board on a solid support that allows it to rotate. Then a small amount of force on the other end of the board will be sufficient to lift the desk. (See Figure 2-7A.) The board itself is a *lever,* and the pivot it rests on is called a *fulcrum.* The point where you apply the force is called the *effort,* which is just another word for input. The effect of applying this force, lifting the desk, is called the *load,* which is the output in this case.

This lever works because the effort moves a lot further than the load. Because the effort end moves a longer distance, it doesn't take as much force as the actual weight of the desk. The load end, on the other hand, moves a very short distance, but with a much greater force. If you moved the fulcrum towards the middle, as in Figure 2-7B, you would lose this advantage. Now the effort end and the load end move about the same amount, and the force is also about the same at both ends.

A lever can make it easier to lift something, or overcome a force of any kind, by multiplying the effect of the force at the effort end. How much does the force get multiplied? This depends on the location of the fulcrum. The closer the fulcrum is to the load end compared with the effort end, the more the effort force is multiplied. These distances from the fulcrum have special names, which are shown in Figure 2-8. The distance from the load to the fulcrum is called the *load arm,* and the length of space between the effort and the fulcrum is the *effort arm.* Their ratio, which tells you how much the effort force is multiplied, is called the *mechanical advantage.* This connection between the ratio of the distances and the ratio of the forces also has a special name. It is called the *Law of the Lever,* and was discovered by Archimedes more than 2000 years ago!

"How Do Levers Make Work Easier?" in Chapter 3 is designed to explore levers and mechanical advantage.

Most people would recognize the desk-lifting board as a lever, but many levers are less obvious. A common idea is that a lever must be straight, but the hammer in Figure 2-9 is also a lever, although it is bent. The handle of the nail clipper in Figure 2-4 is another example of a bent lever. Another common misconception is that the lever must always sit above the fulcrum, but the fulcrum can equally well be on top, as in Figure 2-6B. Figure 2-9 shows the effort, effort arm, load, load arm, and fulcrum for the example of the hammer.

2-7: **Using a level to lift a desk**

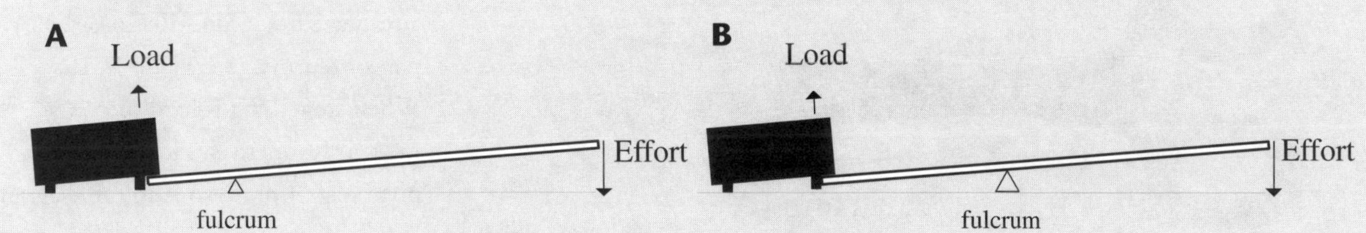

2-8: **The law of the lever**

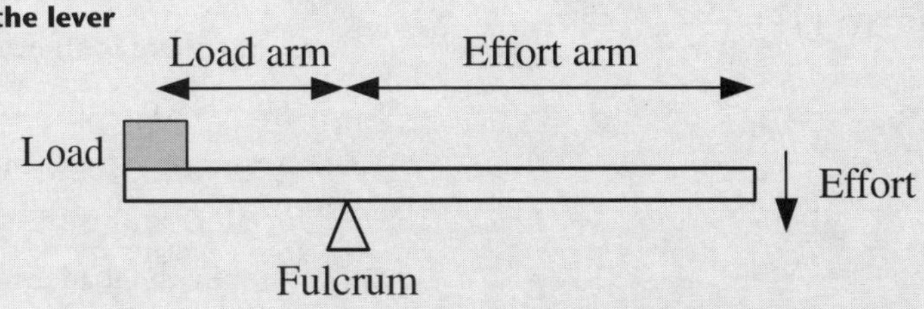

$$\text{Load arm} \times \text{Load weight} = \text{Effort arm} \times \text{Effort force}$$

or

$$\frac{\text{Load}}{\text{Effort}} = \frac{\text{Effort arm}}{\text{Load arm}} = \text{"Mechanical Advantage"}$$

2-9: **A hammer as a lever**

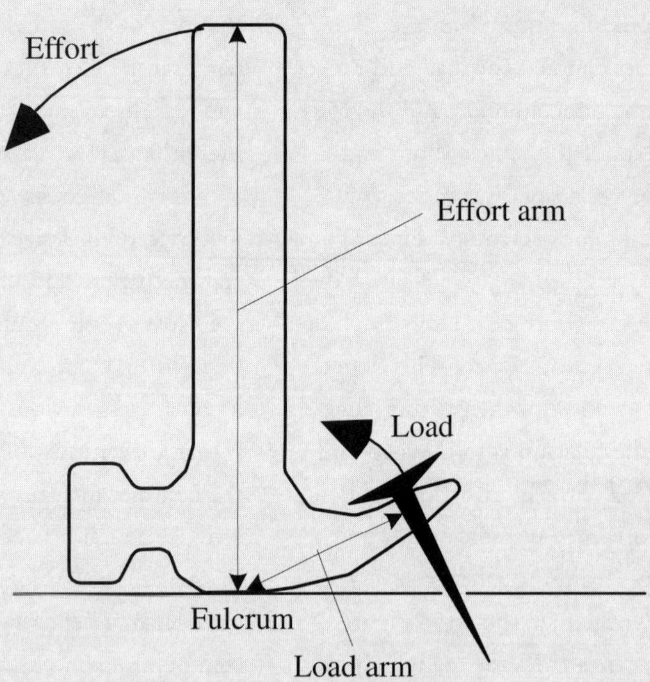

2-10: **Why is it easier to cut something at A than at B?**

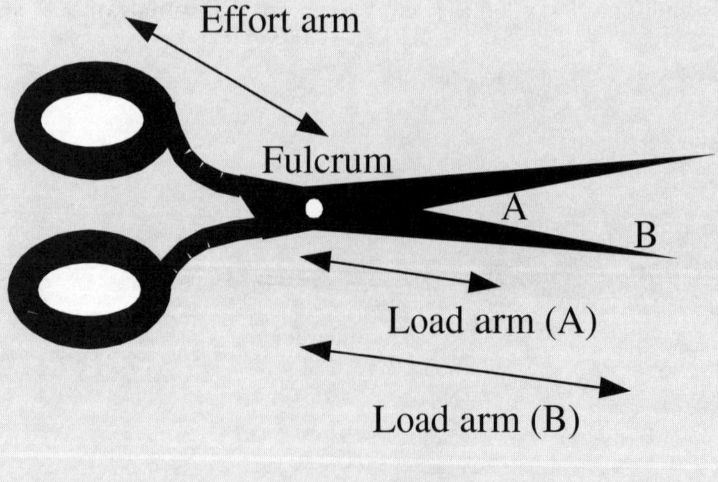

To see the Law of the Lever in practice, you can try the following experiment. Hammer a nail part way into a block of wood, as shown in Figure 2-9. Grab the handle near the end, and pull it towards you. The nail should begin to come out easily. Now try the same experiment with your hand about halfway down the handle. The effort arm is reduced, while the load arm stays the same, meaning that the mechanical advantage is reduced. Sure enough, it is harder to pull the nail out when you move your hand away from the end of the handle and closer to the nail.

An interesting twist with the hammer is that the fulcrum does not stay in one place. As you pull it towards you, you will find the hammer resting on the head. In a way, this works for you, because the mechanical advantage is greatest at the beginning, when you need it the most.

Another example of mechanical advantage has to do with an ordinary pair of scissors. (See Figure 2-10.) When something is hard to cut, you move it instinctively closer to the pivot, e.g., from B to A in the diagram. Why? Most people claim that the blades are sharper there. Perhaps they are, but there is an even better reason to cut closer to the pivot: you are increasing the mechanical advantage. The effort arm is the same in both cases, but the load arm is much shorter when you cut at A than when you do so at B. Therefore the mechanical advantage is greater, which means that the same amount of effort can cut a harder object at A than at B.

Rearranging Levers

The levers discussed so far have all been arranged with the fulcrum between the effort and the load. This scheme of "Effort-Fulcrum-Load" is the most familiar one, and many people contend that the fulcrum is always in the middle. However, two other arrangements are equally possible, namely "Fulcrum-Load-Effort" and "Fulcrum-Effort-Load." Both are very common, and the Law of the Lever applies equally well to them. For example, both a nutcracker and a bottle opener are of the "Fulcrum-Load-Effort" type, or as they're better known, second-class levers. (See Figure 2-11.) The term *second-class lever* implies no value judgment, but is merely a convenient way of distinguishing between this and the more familiar "Effort-Fulcrum-Load" type, which is called a *first-class lever*. Note that a second-class lever always has a mechanical advantage greater than 1, because the load is closer to the fulcrum than the effort is, making the effort arm greater than the load arm (see Figure 2-11B).

If you assumed that the remaining type of lever, "Fulcrum-Effort-Load" should be called a *third-class lever*, you were right! Staple removers and tweezers are both examples of third-class levers. (See Figure 2-12.) As in the second-class lever, the fulcrum is at the end, but in these cases, the effort is closer to the fulcrum than the load. Because the effort arm is shorter than the load arm, the mechanical advantage is less than one, and it takes more force to operate the device than is delivered to the load.

If this is true, why would third-class levers be used at all? Recall that in the example of lifting a desk, there was a price paid for the reduction of effort. Although the effort end didn't need as much force, it had to travel a longer distance. A third-class lever uses this principle in reverse. The effort may require more force, but it doesn't have to go very far. A small movement at the center of the tweezers or staple remover results in a larger movement at the load end, which is why these devices are designed as third-class levers.

Your forearm is another example of a third-class lever. The fulcrum is your elbow joint, the load is your hand and whatever it happens to be lifting,

2-11: **Two second-class levers**

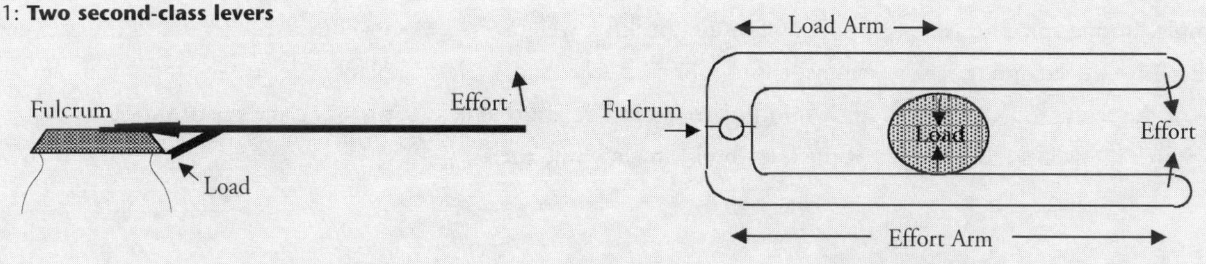

A: Bottle Opener **B. Nut Cracker**

2-12: **Two third-class levers**

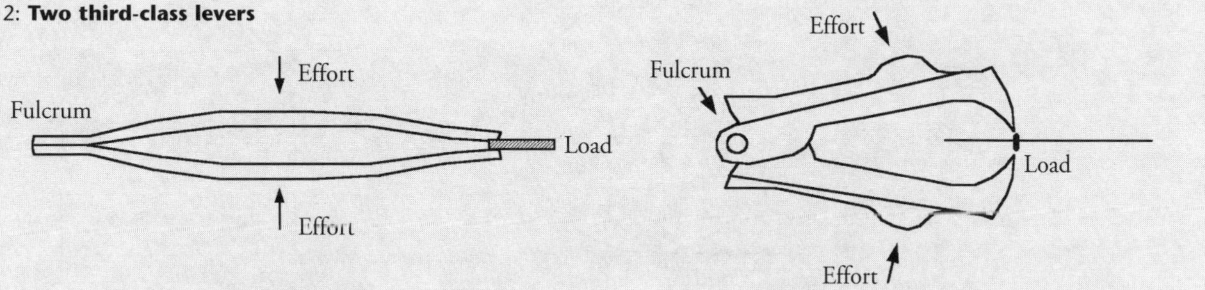

A: Tweezers **B. Staple Remover**

Table 2-2
THREE CLASSES OF LEVERS

Type	Description	Mechanical Advantage	Examples
1st Class	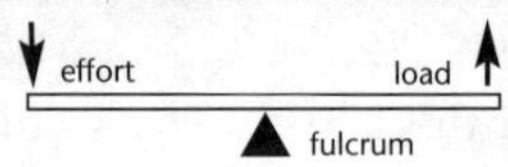	less, greater or equal to one	scissors, pliers, oars, hammers (as nail extractors), can opener, hand truck
2nd Class	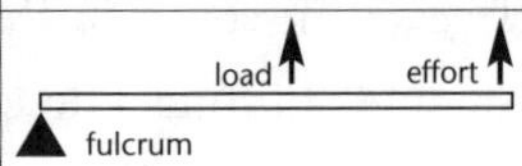	greater than one	wheelbarrow, nutcracker, garlic press, bottle opener, handlebars
3rd Class	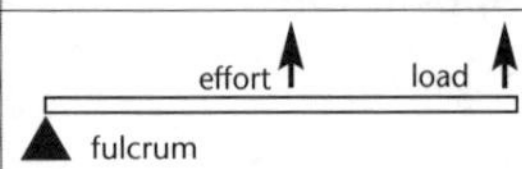	less than one	forearm, tweezers, staple remover, fishing rod, shovel

and the effort is supplied by a muscle which connects the upper arm to the lower arm just inside the joint. The arms of cranes and bulldozers work the same way. Table 2-2 summarizes the three classes of levers.

Science books usually identify six *simple machines: the lever, wheel-and-axle, pulley, inclined plane, wedge,* and *screw.* Two of these—the wheel-and-axle and the pulley—are really examples of the lever.

Figure 2-13 shows a wheel-and-axle, for example, from a car. The axle makes the wheel turn, overcoming the resistance of the road. To make the vehicle go forward, the wheel has to push back against the road, which is the load in this case. The effort is supplied by the axle, which is much closer to the fulcrum (at the center) than the road is. Therefore, a wheel-and-axle is a third-class lever. The dashed line shows what an ordinary straight lever would look like if it were doing the same job as the wheel-and-axle. Of course, the lever would only work for an instant, because as soon as it rotated, it would no longer be in contact with the road.

A pulley, used to lift a weight, is equivalent to a first-class lever, as shown in Figure 2-14. The effort side of the lever pulls down, while the load side lifts the weight up, and the fulcrum is at the middle. The dashed line shows how the pulley could be replaced momentarily by a straight first-class lever, which would do the same job.

Two of the other three simple machines—the wedge and the screw—are examples of a third kind of simple machine—the inclined plane. The lever and the inclined plane are basic to all mechanisms. It would be more accurate to say there are only two simple machines, the lever and the inclined plane, because the other four are really just special cases of these!

2-13: **The wheel-and-axle as a third-class lever**

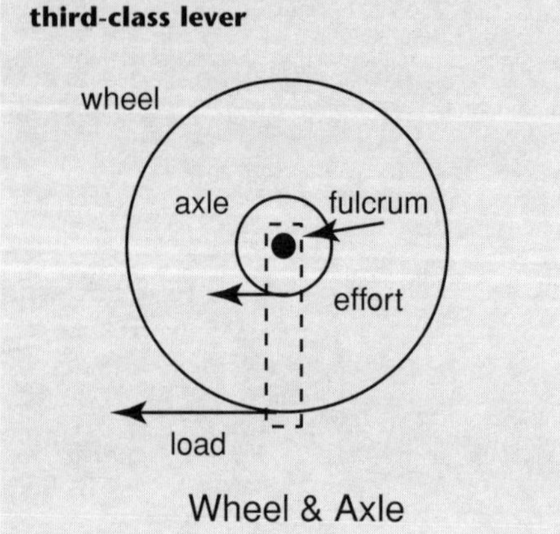

2-14: **The pulley as a first-class lever**

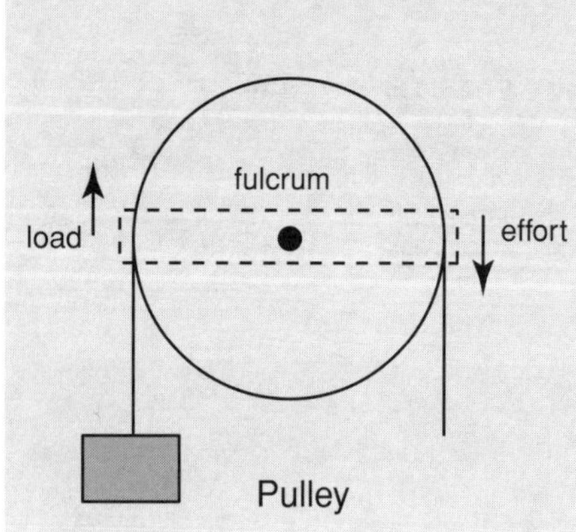

Compound Levers and Linkages

Most of the examples so far show individual levers operated by hand, as in the cases of the hammer and the bottle opener. In some cases, like the tweezers, the scissors, and the staple remover, two identical levers use a common fulcrum and are hand-operated in tandem. Each of these devices could be described as a double lever. In the case of the nail clipper, shown again in Figure 2-15, we see a system of two levers arranged so that one lever operates another. The handle is a second-class lever (A), whose load (output) is the effort (input) to the upper jaw, which is a third-class lever (B).

A device like the nail clipper, in which one lever acts on another, is called a *compound lever* or *linkage.* Other examples of linkages include folding chairs and baby carriages, vise-grip pliers, umbrellas, lawn sprinklers, windshield wiper mechanisms, adjustable-arm desk lamps, clothes drying racks, "lazy tongs" lamps and mirrors, "pop-out" tool boxes and sewing boxes, bicycle handbrake mechanisms, manual typewriter mechanisms, and most automotive hood and tail-gate opening hinges.

2-15: **The nail clipper as a compound lever or linkage**

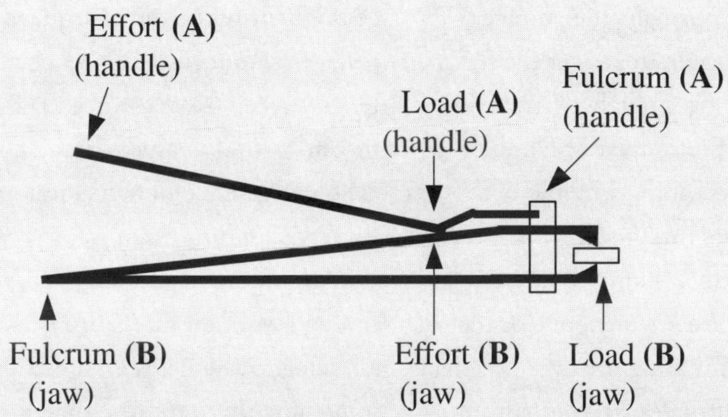

How Children Understand Mechanisms

Although little research has been done about children's conceptions of mechanisms, Piaget did look at this issue. In *Success and Understanding,* he reports on research regarding levers. Figure 2-16 shows one example.

A child of nearly five years was presented with a hotizontal first-class lever. The fulcrum was in the center of the bar. The problem was: How can you lower a block of sugar at Y using your finger at X? At first, the child simply lowered the block by

2-16: **A five-year old's view of a first-class lever**

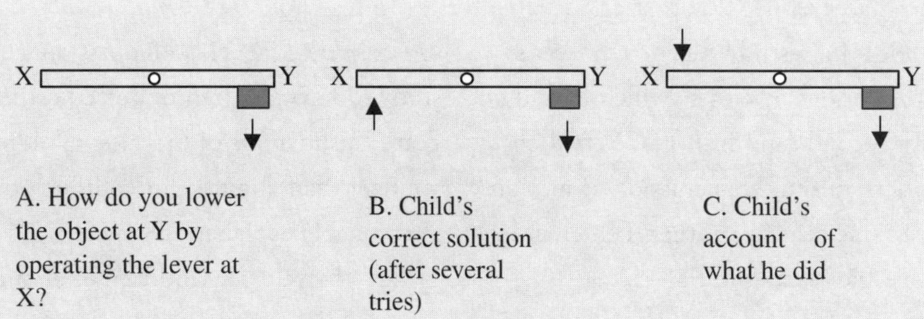

hand, without using the bar. Then he claimed there was no way to do it with the bar, because the bar wouldn't budge when he pulled it down at the pivot. Then, with a little coaching, he managed to lower the block at Y by raising the bar at X. When asked how he had done it, he insisted that he had pushed the bar down at X, rather than up. Apparently, he could not imagine that the bar could actually go up at one end and down at the other!

Piaget explains that this sequence reflects a lag between the child's intuitive hands-on knowledge and his ability to conceptualize it abstractly. In other words, a young child can perform the task without being able to explain it correctly. According to Piaget, as children mature, conceptualization gradually becomes more important. By early adolescence, children actually form a concept of the situation before taking any action.

Piaget's research occurred in the context-free setting of a pegboard lever on a pegboard base. Presumably, the children he studied had no prior experience with this type of apparatus, which was not part of any mechanism they had ever used. Lehrer & Schauble (1998) did a study called "Children's Conceptions of Gears" which looked at how second- and fifth-grade children understand the transmission of motion from one gear to another. Lehrer and Schauble asked questions like:

1. If one gear is driven, what makes the other one turn?
2. Do the two gears rotate in the same or different directions?
3. What determines the relative speeds of the two gears?

Some of this research was done using gears on a pegboard isolated from the rest of the world. The same questions were later asked of children examining an eggbeater and a ten-speed bike. Not surprisingly, the younger children were able to answer the first question more accurately when it was presented in the context of familiar devices. In describing why the both pegboard gears turn at the same time, one child said, "It's kind of like a copycat." In other words, one of the gears is just imitating the other. He had no concept of the causal mechanism linking the two gears. In looking at the eggbeater, the second-graders were much more likely to recognize the role of the gear teeth in transmitting the motion from one gear to the next. Lehrer & Schauble's work underscores the importance of teaching mechanisms using artifacts that are already familiar to children.

In his book, *The Child's Conception of Causality*, Piaget also reports on a study of how children understand the gear transmissions of bicycles. Children younger than about eight couldn't offer a reasonable explanation of what makes a bicycle go. Children of about four and five gave accounts that did not involve the feet or the pedals at all. Their explanations included "it must go," "the lamp," "the street," "the air in the tires," etc. At about six or seven, children identified the parts involved in making a bicycle go, but did not understand their cause-and-effect relationships. One child, for example, saw that the pedal operates the front sprocket, but thought that the rear wheel drove the chain! Starting at about eight years, Piaget's subjects could identify the causal sequence pedal –> front sprocket –> chain –> rear sprocket –> rear wheel. Perhaps the most striking aspect of these studies is how different children's ideas are from what adults assume.

An important idea that is developed through work on mechanisms is modeling. A model is an imitation of something that captures some fundamental features, but is also different from the real thing. Chapter 1 discusses how to make and troubleshoot models of mechanisms. Penner, et al, (1997) describes how first- and second-graders grappled with the task of making functional models of the human elbow. In the course of this work, they gradually came to distinguish between models that look like the real thing and those that work like the real thing. Based on this work, the children became much more knowledgeable both about simple mechanisms and also about the nature of models.

Mystery Mechanisms: The Design Puzzles Solved!

We will conclude our discussion of the concepts underlying the study of mechanisms by solving the mechanism design problems posed at the end of Chapter 1, shown in Figures 1-42, 1-43, and 1-44.

2-17: **Design Puzzle #1**

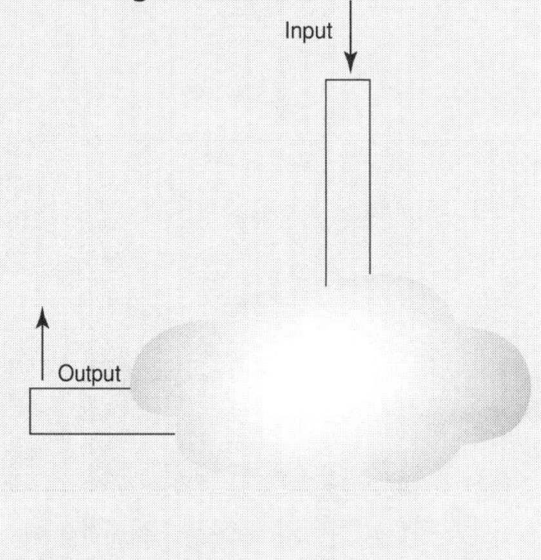

Design Puzzle #1: The Arms Flap Up When the Head Goes Down!

The problem is shown in Figure 2-17. To solve this problem, think about what needs to happen at the output. It will need to swing upward, as a result of an input that pushes downward. What sort of lever has opposite directions of motion at the input and output? Looking at the scissors in Figure 2-10 or the desk-lifter in Figure 2-7, it is clear that a first-class lever will do this job. Figure 2-17 is redrawn as Figure 2-18, showing the output link as a first-class lever. The solid circle indicates that the fulcrum is a *fixed pivot* attached to the base of the mechanism.

Now what could be behind the remaining cloud? Something is needed to transmit the vertical translation from the input to the right-hand side of the first-class lever. Clearly, a slider will accomplish this task, and the complete design is shown in Figure 2-19. The connection to the first class lever is made by a *floating pivot* represented by the open circle. A floating pivot connects two links, but does not attach them to the base. Note that the sliding link is constrained by a guide (see Figure 1-40). The circles represent paper fasteners.

The width and location of the guide are important. If it is too tight or too far from the floating pivot, the link may not be able to rotate slightly, allowing it to turn the output link. If, on the other hand, the guide is too loose, it may have too much side-to-side movement. Another thing you can play with is the location of the fixed pivot on the output link. If it is moved to the right, the output motion will be greater, but more force will be required at the input. The opposite will be true if it is moved to the left. As with any design problem, a great deal can be learned from trying to rework the design once you have a first working model.

2-18: **Using a first-class lever**

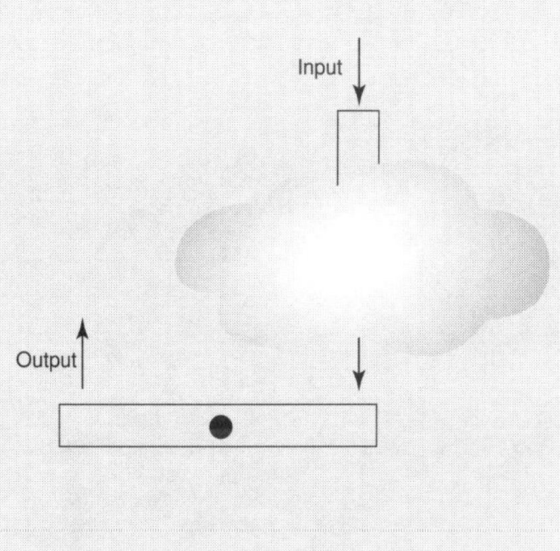

2-19: **Solution to Design Puzzle #1 showing slider**

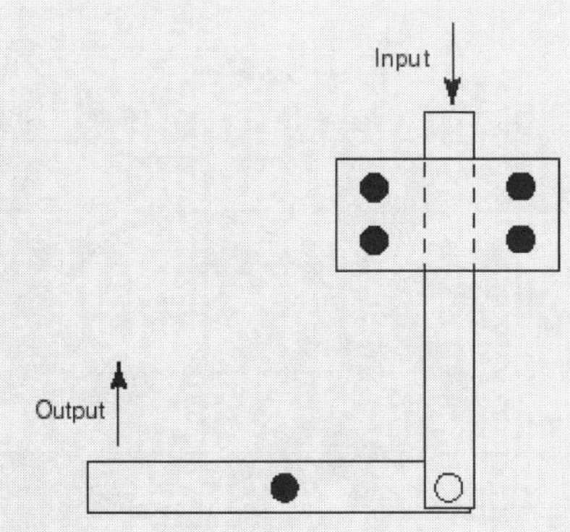

Design Puzzle #2: The Arms Shoot Out When the Head Goes Down!

In this problem, the arms shoot out instead of up when the head is pushed down. As in the first problem, we start with one side only. The input and output are shown in Fig. 2-20.

Although apparently similar to Design Puzzle #1, this challenge is actually quite different. In Puzzle #1, the direction of motion had to change by 180°. Here, a 90° change of direction is needed between input and output. None of the three classes of levers shown in Table 2-2 will accomplish this result, but the hammer in Figure 2-9 offers a clue. Although the hammer is a first-class lever, the effort and load are different by much less than 180°. The reason is that the hammer is a bent lever—the bar itself changes direction.

Another way to see how to make the change of direction is to make an analogy between the lever and the wheel. We have already seen, in Figure 2-14, how the opposite sides of a pulley behave like a first-class lever. Figure 2-21 extends this analogy by showing two other ways to make a wheel act like a lever. Figure 2-21(A) restates the idea of Figure 2-14, showing a wheel as a straight first-class lever, with the effort and load at opposite ends and moving in opposite directions. Figure 2-21(B) is a simplified view of the hammer in Figure 2-9, and an equivalent circle. This time, the input and output are not taken from opposite sides, but only about 60° apart. Figure 2-21(C), finally, offers a solution to the

2-20: **Design Puzzle #2**

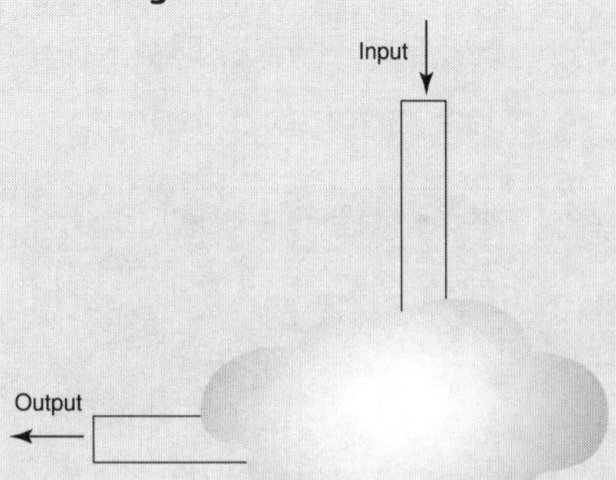

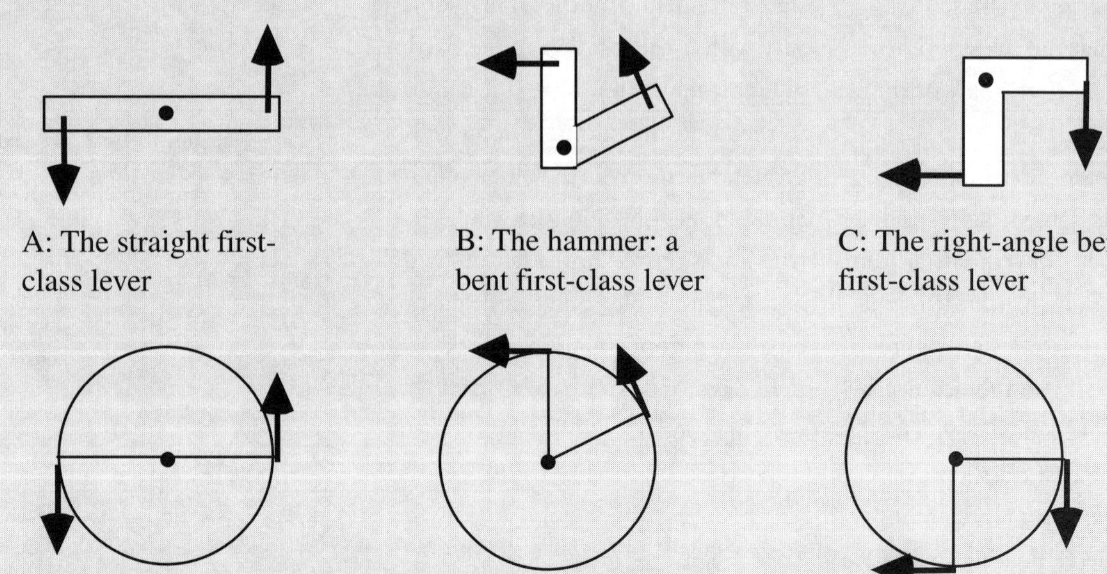

2-21: **Three analogies between wheels and levers**

right angle problem of Design Puzzle #2. It shows how both a wheel and a bent lever can change the direction between the effort and the load by 90°.

An important thing to note in all three sets of diagrams is that you don't need the entire wheel to do the job. In Figure 2-21(A), a straight bar replaces the wheel. In terms of input and output, they both do the same thing. In Figure 2-21(B) and (C), the wheels are also replaced by bars, but in these cases the bars are bent. The bent bars work the same way as the wheels except that they are not as strong. To make the L-shaped lever stronger, all you have to do is fill it in; a triangle is much stronger than an L. You could also use a quarter circle, which is similar in shape to a triangle. The rest of the circle is simply not needed. These shapes are compared in Figure 2-22.

Using the ideas from Figure 2-21(C) and Figure 2-22, it is easy to design the basic mechanism for Design Puzzle #2. Figure 2-23 shows how to use a triangle to change the direction of motion by 90°.

2-22: **From circle to bent lever**

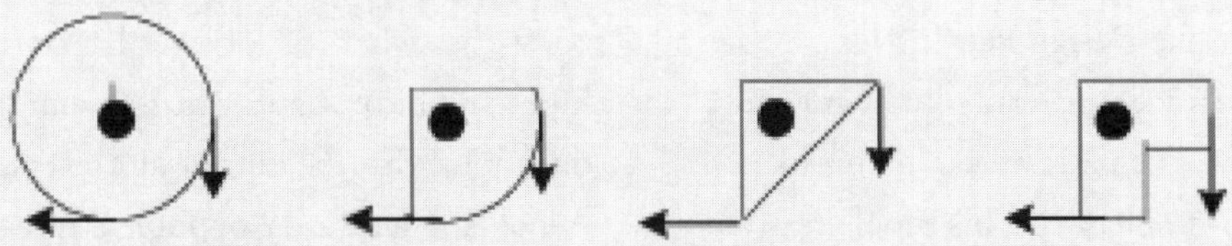

2-23: **Using a triangle to make a bent lever**

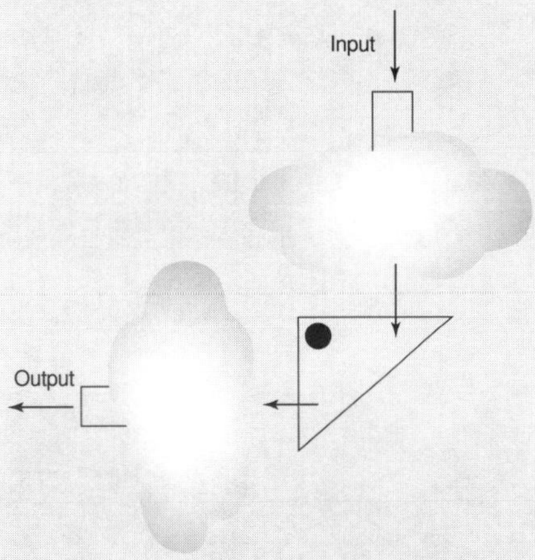

2-24: **Solution to Design Puzzle #2**

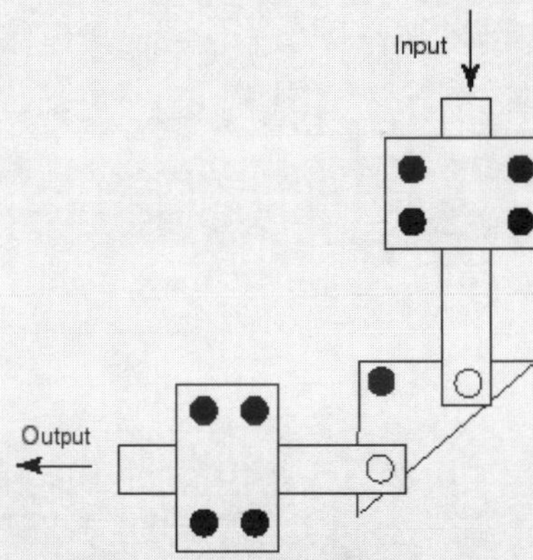

The rest of the problem consists of transmitting the forces from the input to the bent lever, and from the bent lever to the output. This is accomplished, as in the previous problem, by adding sliders. A complete solution is shown in Figure 2-24. Note that the sliders are attached to the triangle by floating pivots, while an off-center fixed pivot attaches the triangle to the base.

Design Puzzle #3: The "Kissing Couple" Problem

2-25: **Design Puzzle #3**

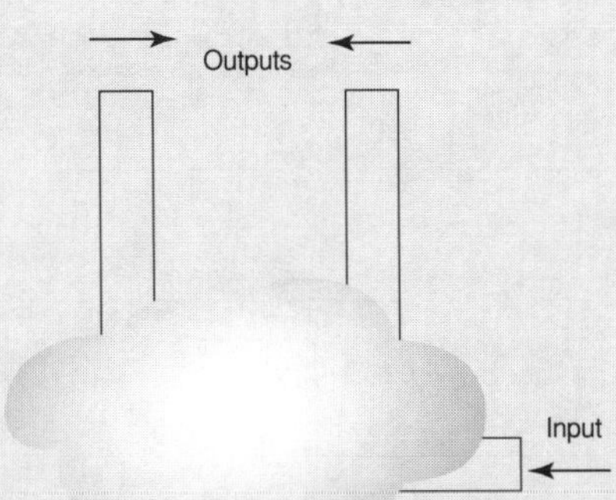

The problem in Design Puzzle #3, shown in Figure 2-25, was to make a little toy that has an input on the side which is pushed in, and two outputs on the top which come together.

The first thing to notice about this problem is that it really consists of two problems, each having one input and one output. The input is the same for both problems. These are shown in Figure 2-26. Breaking the problem up into two little problems is an example of the "divide-and-conquer" strategy for solving problems.

Sub-problem A is actually the same problem as Design Puzzle #1 (arms wave up when the head goes down) except that it has been turned on its side. The equivalence between these problems is shown in Figure 2-27.

If you turn the book 90° counter clockwise, sub-problem A will look just like Design Puzzle #1.

Since we already know how to solve this part, we can go on to sub-problem B. This problem is very similar to Design Puzzle #1, but with one very important difference: the output and the input go in the same direction rather than opposite directions. So, we ask what sort of lever has both input and output going in the same direction, with the output at one end. A look at Figure 2-12 or Table 2 reveals the answer: a third-class lever will work. Figure 2-28 shows a solution to sub-problem B with a third-class lever.

The only remaining step is to combine the solutions of the two sub-problems into the solution for the entire problem. This is done by using the same input for both outputs, shown in Figure 2-29.

2-26: **Design Puzzle #3 broken into two sub-problems**

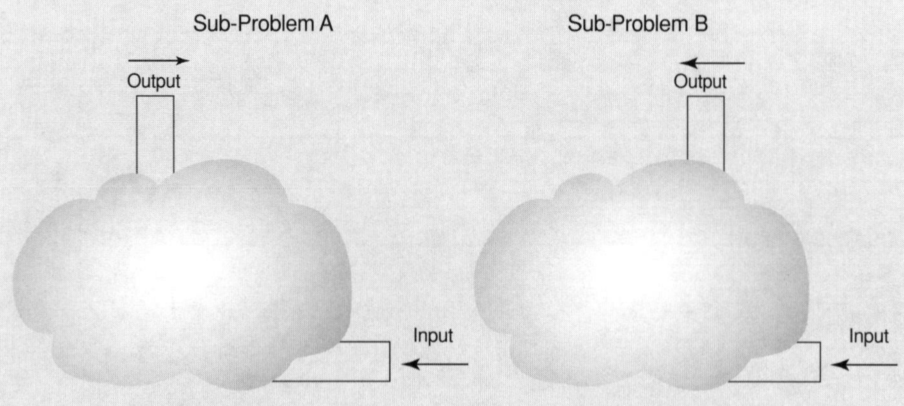

2-27: **Comparing sub-problem A of Design Puzzle #3 with Design Puzzle #1**

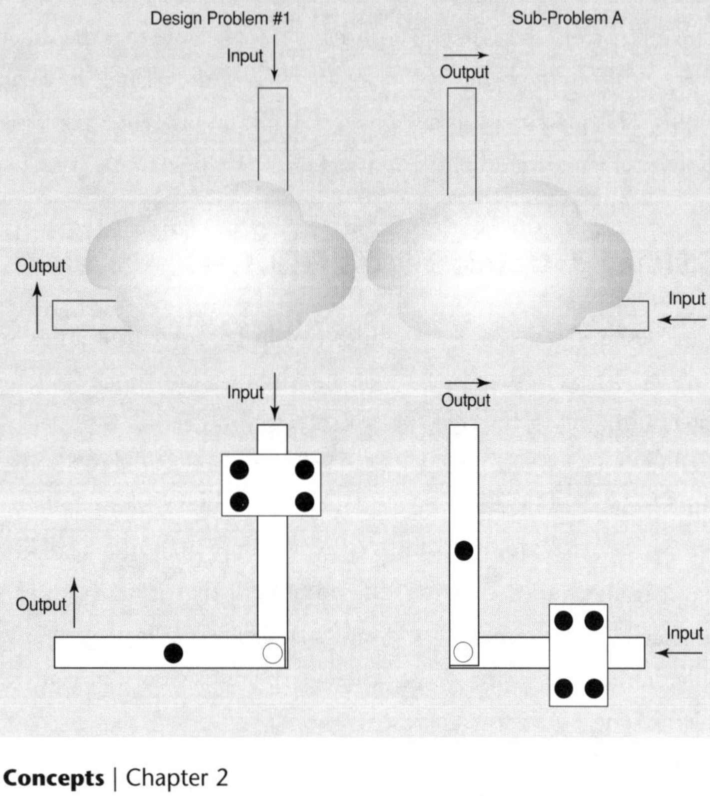

2-28: **Sub-problem B of Design Puzzle 3 is solved.**

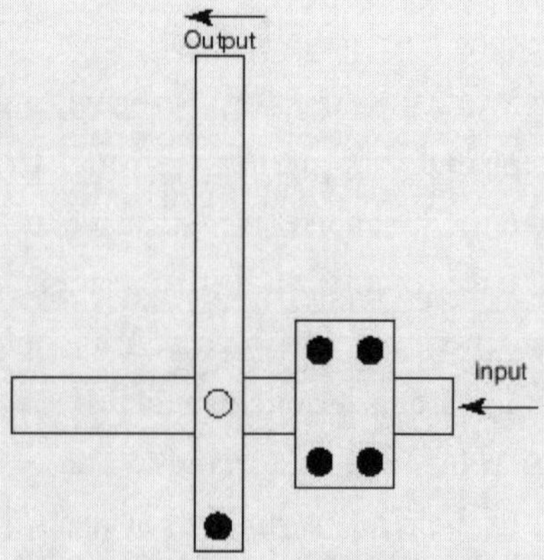

2-29: **Solution to Design Puzzle #3**

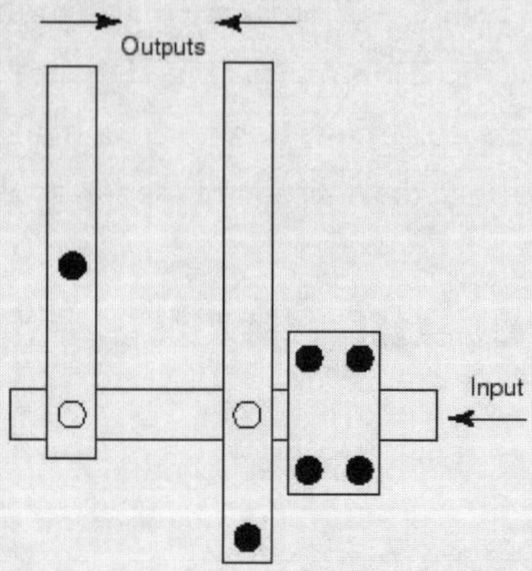

How Circuits Work

The Flashlight, Revisited

In Chapter 1, "Appetizers," we raised the question of how a flashlight works. We saw how the switch completes a circuit consisting of the two batteries, bulb, switch, metal strip, and metal spring. When the circuit is complete, the bulb should come on. In a way, this discussion raised a lot more questions than it answered:

- Why doesn't the bulb light when the switch is OFF, considering that the batteries are still in contact with the tip of the bulb?
- Why doesn't the current flow through the air? Why does it seem to flow only through the metal parts?
- What do the batteries do?
- How come the bulb lights up when a current passes through it?
- Which part actually produces the light?
- Why is there a glass bulb around the whole thing?
- What happens when a bulb "blows"?

Let's start with some basic concepts of electricity. All electrical phenomena are explained by the existence of *charges.* These are invisible but very real properties that come in two varieties, which the early discoverers, such as Benjamin Franklin, arbitrarily called "positive" and "negative." It later turned out that electricity is mostly carried by *electrons,* which carry the negative charges. The important thing is that the flow of electricity requires that these little particles move from one place to another. This idea is suggested by words like *flow* and *current,* which make the analogy between electricity and the movement of fluids such as water.

2-30: **Typical dry cell battery**

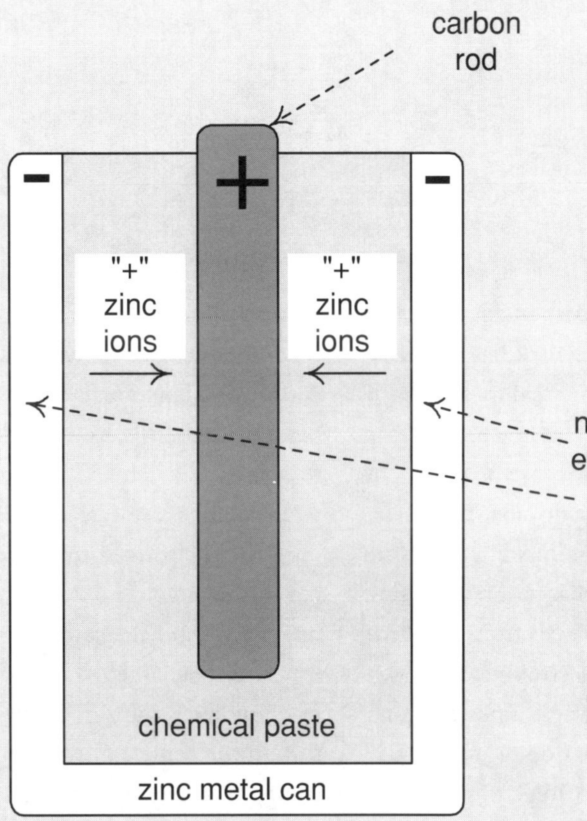

Now, here is a little problem: What incentive is there to make these little negatively-charged electrons move? The answer is contained in the observation that like charges repel while opposite charges attract. This implies that if you have a bunch of electrons in one place, because they all have negative charges, they will try to get as far apart from one another as possible. On the other hand, if you can create a positive charge somewhere, electrons will be attracted there. How can you do this?

2-31: **Battery in a simple circuit**

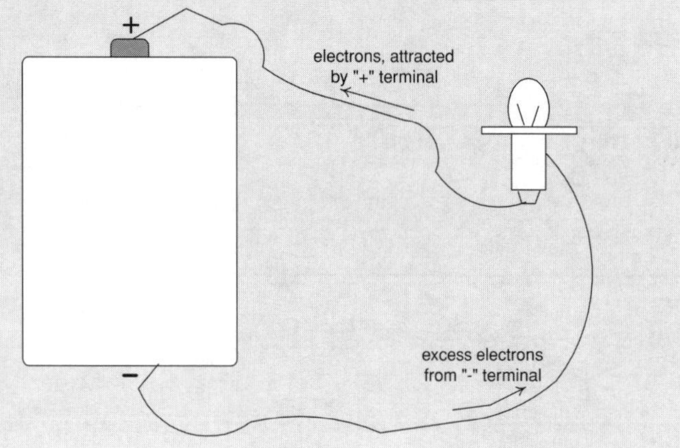

2-32: **The parts of a small lightbulb**

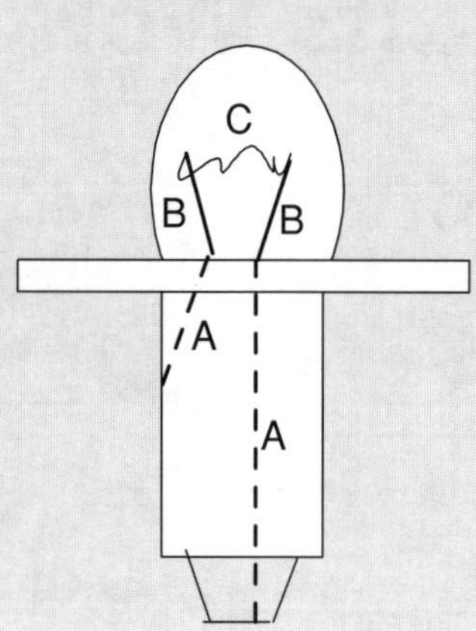

Every *atom* is *neutral*—it has just as many positive charges, which are contained in its nucleus, as it has negative charges, which are carried by the circulating electrons. If an atom loses one of these electrons, the positive charges will now outnumber the negatives, and the atom will be left with an overall positive charge. An atom that has lost one or more electrons is called a positive *ion.* An electron nearby will be attracted towards a positive ion, because of the attraction of unlike charges. So, the trick in getting electrons to flow is to somehow remove electrons from a whole bunch of atoms.

So, to make a current flow, there has to be some special way of pulling electrons off of atoms and overcome the tendency of these atoms to hold on to the electrons. This situation is somewhat like rolling a rock up a hill. Obviously, the rock would "rather" stay at the bottom, but if you exert enough energy, you can roll it all the way to the top. In an electric circuit, the thing that "gets the rock up the hill"—i.e., pulls the electrons off the atoms—is the battery.

Like food or gasoline, the battery comes with energy stored in its chemical components. The cheapest batteries consist of a zinc metal can, a gooey paste of chemicals, and a carbon rod in the center. The chemicals attract positive zinc ions away from the can, leaving behind electrons that give the can a negative charge. The positive zinc ions migrate toward the carbon rod, making it positively charged. The chemical reaction keeps the zinc ions flowing toward the center. This is how chemical energy is converted into electrical energy, as shown in Figure 2-30.

Now, suppose we attach a light bulb to the battery, via two metal wires, as shown in Figure 2-31. Now, the excess electrons on the negative terminal of the battery, left behind by the departing zinc ions, have somewhere to go. Attracted by the battery's positive terminal, they flow through the bulb and back to the battery.

Why did we say that the wires had to be made of metal? Electrons can travel much more easily in some materials than in others. In a metal, there are a lot of electrons that are free to move fairly easily. These free-floating electrons account for all of the major characteristics of metals: they conduct heat, reflect light, and conduct electricity.

Next, let's see what happens in the bulb. A little bulb works the same way as an ordinary domestic light bulb. The only part that actually lights up is a thin wire filament made of tungsten, a relatively rare metal. (See Figure 2-32.) This filament (C) is supported by the two visible wires (B), which attach to the tip and the side of the base by some hidden wires (A).

Why does the filament light up, but not the other wires? The filament is a wire too, but it is so thin that it simply doesn't have enough room to carry all of the electrons easily. As a result, there are many more collisions within the filament than within a normal wire. These collisions heat the filament up until it glows, which accounts for the light. The glass bulb serves to keep oxygen out. If oxygen were present while the filament was white hot, the filament would simply burn up. You can see this happen occasionally with a domestic light bulb that hasn't been used for a while. Over time, air has leaked into the bulb, and the next time you turn it on, POOF!!! It glows very bright for an instant, and then blows.

Something else that can go wrong with this circuit is that the battery can go "dead." This happens when the chemical paste loses its power to push zinc ions away from the can and towards the carbon rod. Then there is no longer a mechanism for making the electrons flow in the circuit.

Now, suppose we add a switch to this circuit. You can make a crude switch just by disconnecting one of the wires and alternately touching it and removing it from the metal case of the bulb. (See Figure 2-33.) Why does the light go off when the wire is no longer touching? When the wire is removed, the electrons would have to flow through the air in order to complete the circuit. To make electrons flow through air requires a lot more energy than is available, because nearly all the electrons in air are tightly bound to atoms. The same is true of all the other materials known as *insulators.*

The makeshift switch that is made by disconnecting and reconnecting the wire does exactly the same thing as a manufactured switch, except that manufactured ones are more reliable. All they do is disconnect and reconnect the wires when you move the lever, push the button, turn the knob, or slide the button. The configurations in Figures 2-31 and 2-33 work the same way as the flashlight in its ON and OFF positions, respectively. In both cases, you have a battery, bulb, and metal pieces connecting them. The major differences are:

1. The flashlight has two batteries instead of one, and a more reliable switch.
2. In the flashlight, one of the batteries is directly in contact with the bulb, with no wire in between.
3. In the flashlight, there is a metal spring and metal strip connecting the battery to the other side of the bulb, instead of a wire.

From the point of view of the electrons, which do all the work in this circuit, none of these differences matters very much, except that two batteries give them twice as much energy as one. The circuit in the diagram is otherwise equivalent electrically to the flashlight circuit, although they look very different.

One way to show the equivalence between the circuits is to use a special kind of map called a *schematic diagram* to represent them both. A schematic uses a standard set of agreed-upon symbols to represent all of the components. Schematic diagrams are a bit like musical scores, in that all of the people who use them have learned to interpret the symbols.

Schematic circuit symbols save a lot of time and effort, because they leave out all of the details that are unimportant to the functioning of the circuit, such as the colors of the wires or how long they are. A schematic diagram of the flashlight circuit is shown in Figure 2-34. The standard circuit symbols are explained in the key.

2-33: **Simulating a switch by disconnecting a wire**

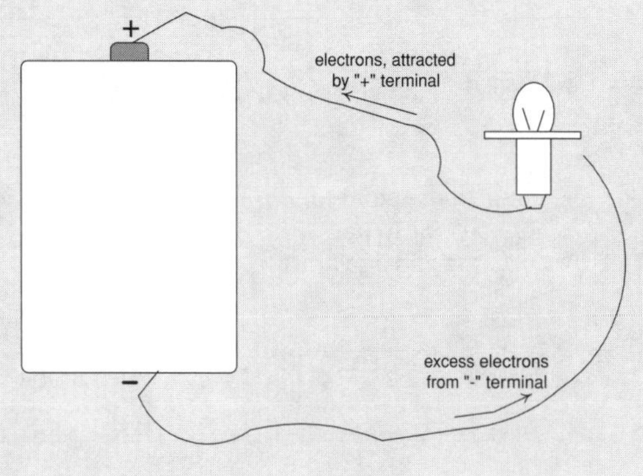

2-34: **Schematic diagram of flashlight circuit**

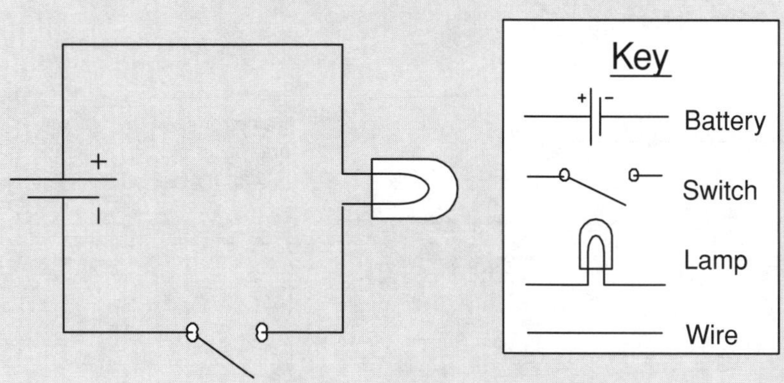

Circuit Concepts and Popular Beliefs

The picture we have presented so far of a basic circuit says that when the bulb lights, the following things happen:

1. Current, in the form of moving electrons, flows from one side of the battery, through the bulb, and back to the battery;
2. The battery supplies energy to the electrons, which the bulb converts to light and heat energy; and
3. The same amount of current flows all the way around the circuit, because the electrons eventually return to the battery to be energized again, as illustrated in Figure 2-35.

Although these are the basic ideas accepted by scientists, popular beliefs about circuits are quite different. These alternative ideas begin in childhood, as Tasker and Osborne (1985) show, and are still held by students who have passed college physics, as McDermott & Schaffer (1992) demonstrate. Few adults seem to develop the concepts accepted by modern science. According to this research, there are three basic "folk theories" of electric circuits:

1. The "single-wire" theory says that you really only need one wire to light the bulb. When confronted with the evidence that the bulb won't light, until the second wire is attached, children have claimed that the other wire is "just an extra" or "a safety wire." They see the battery as a thing that "gives" electricity, much as a faucet provides water. This theory may be a by-product of the analogy between water and electricity. (See next section.) This theory is shown in diagram form in Figure 2-36.
2. The "clash-of-currents" theory accepts that both wires are needed. It claims, however, that the current flow in each wire is toward the bulb. When the two currents reach the bulb from

2-35: **Circuit operation, summarized**

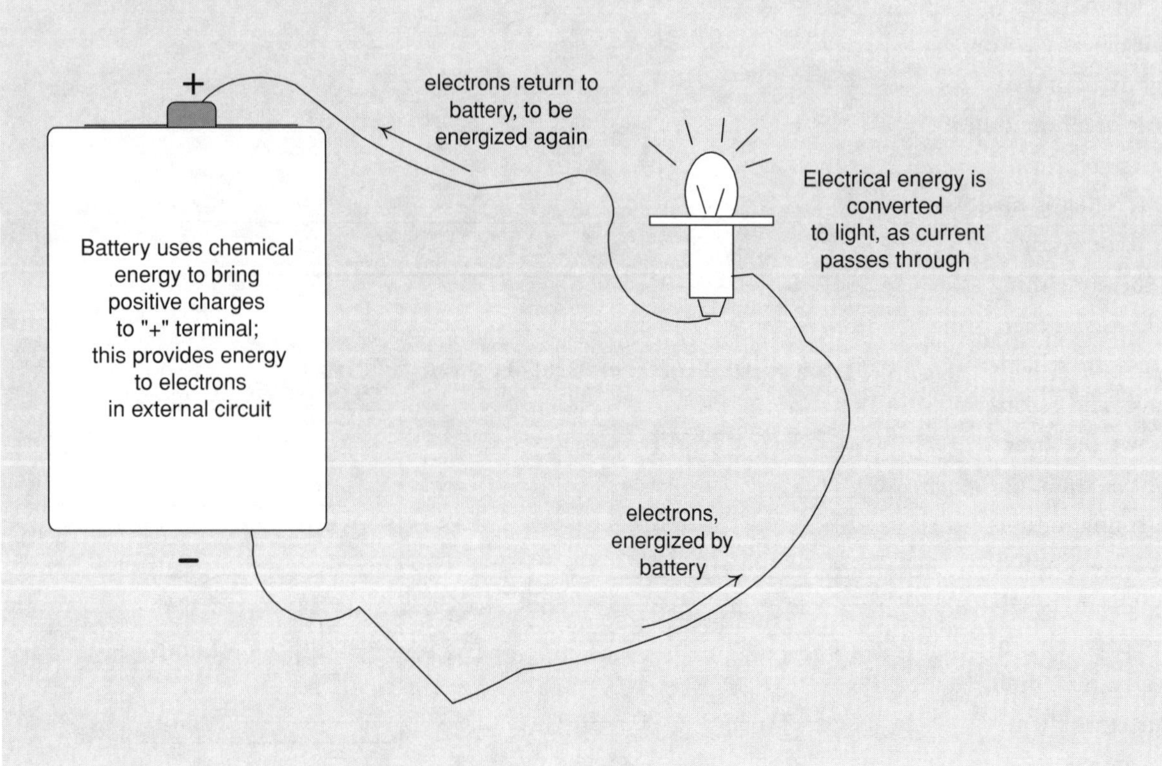

2-36: **The "single-wire" theory**

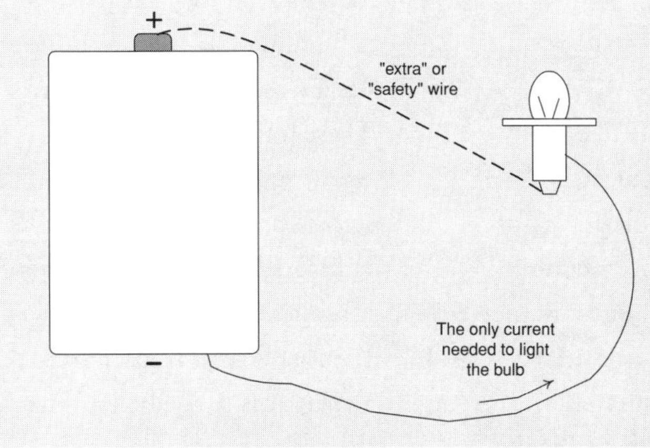

2-37: **The "clash of currents" theory**

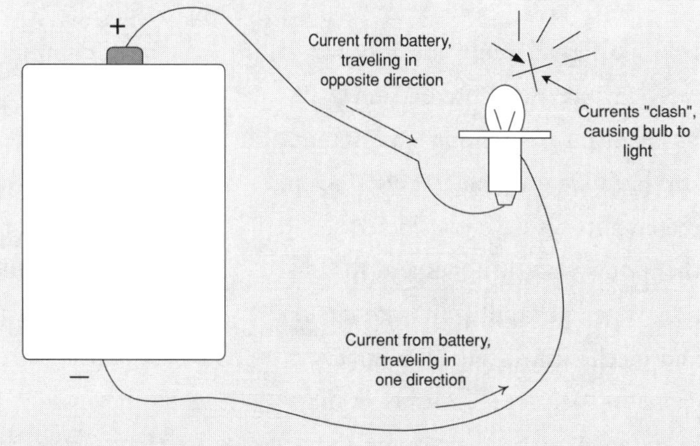

opposite directions, they "clash," producing light. This idea is depicted in Figure 2-37.

3. The "destroyed current" theory says that the current leaves the bulb in the same direction that it came in, but says that some of this current gets "lost" or "destroyed" in the bulb. As a result, there is less current leaving the bulb than there was coming in. This theory identifies current, rather than energy, as the property of electricity that is converted into light. Therefore, some of the current gets "used up" in the bulb, and the current leaving is much less than that entering. A diagram is shown in Figure 2-38.

2-38: **The "destroyed-current" theory**

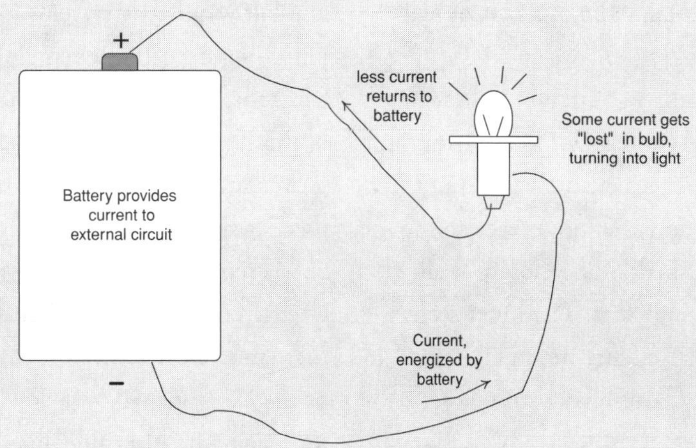

The research on popular concepts of electricity was designed to find out how people understand science concepts such as current and voltage. It involved circuits with batteries and bulbs only. Nearly everyone's daily experience with circuits involves turning them on and off with switches, but switches are not part of this research. Here is an example of the difference between a science orientation and a technology approach. Science education tries to teach the underlying principles of circuit operation, while technology education focuses on the uses of circuits in everyday life.

In our own work with electric circuits, we have found that prior knowledge of science concepts does not necessarily translate into an understanding of technology, and vice versa. Some teachers who were proficient with science units on "Batteries and Bulbs" had difficulty adding switches to their circuits or seeing the relationship between circuits in the laboratory and those in their homes. At the same time, other teachers with little or no experience in science could figure out quickly how to make simple circuits with switches and see how these related to their own experiences. Science and technology look at the same artifacts in different ways, and the kinds of knowledge they emphasize are different.

Circuits, Mechanisms, and Analogies

In some ways, circuits are a lot like mechanisms. Both circuits and mechanisms have inputs and outputs and both are examples of systems. In the circuits we have considered so far, the inputs were switches and the outputs were light bulbs. In both circuits and mechanisms, there is a direct cause-and-effect relationship leading from the inputs to the outputs: throwing the switch makes the bulb light and operating the handle of a pair of nail clippers makes the jaws come together.

Unfortunately, circuits are not nearly as easy to explain as mechanisms. In a simple mechanism, you can usually figure out how the input leads to the output just by looking closely at all of the moving parts. The inner secrets of an electric circuit are much harder to fathom, because you can neither see nor touch the tiny electrons that are doing all the work. A simple mechanism is more-or-less transparent; you can take it apart and see for yourself how it works.

Because you can't see inside circuits, some analogies have been developed to make them easier to understand. These analogies are statements that "a circuit really works a whole lot like such-and-such," where "such-and-such" is something that you *can* see and/or touch.

The most commonly used analogy relates the flow of electricity to the flow of water. The wires are like pipes, a switch is like a valve or faucet, and a battery is like a pump. The energy of flowing water could be converted to a useful form—for example, into turning a water wheel or turbine. This device would be analogous to a light bulb.

However, there is a major difficulty with this water analogy. The most familiar water flow systems involve plumbing fixtures such as sinks, toilets, bathtubs, and washing machines. All of these are *open systems,* which do not recirculate the water once it has been used. It simply goes down the drain and is replaced by new tap water. In an electric circuit, on the other hand, there must be a complete, closed path or the electrons will not flow anywhere in the circuit. After they have lost their energy, for example in a light bulb, they must be returned to the battery to make their journey again. Most circuits are *closed systems* where the "electrical fluid" is recirculated.

In order to make a better analogy, we need to think of a closed system that keeps the same fluid flowing continuously in a circuit. Here are two examples of closed fluid flow systems:

- A building that uses steam heat has a closed system of pipes. These lead from the boiler, which turns water to steam. The steam passes through all of the radiators in the building, where it heats the indoor spaces. Eventually it turns back to water and returns to the boiler, where the cycle starts over.
- A car radiator cools the water/antifreeze mixture that carries heat away from the engine. This fluid is forced through the engine by the water pump, and then back to the radiator where it gets cooled again.

In both systems, a fluid is forced through the system by a boiler or pump that gives it energy, playing the same role as a battery in a circuit. As it flows, it heats or cools the environment, which is analogous to the electrons making the bulb light up. Finally it returns to the original heater or pump, and begins the cycle again. Each of these systems is a closed circuit, like an electric circuit, in that the same stuff flows around and around. The problem is that these systems are also closed in the sense that you can't see what's going on inside them!

To visualize an electric circuit, we need to make an analogy with some kind of circuit that is open to view. One possible analogy is with a roller coaster. At the bottom of a hill, the cars are attached to a tractor, which pulls the cars up a big hill. At the top of the hill, the cars are released, and they coast down to the bottom. The energy they have gained sends them up another small hill. Eventually, they return to the foot of the big hill, and the tractor does its work again. The tracks of the roller coaster are similar to the wires in a circuit, and the cars are like the electrons. The tractor is like the battery. When they go up a little hill, from the energy gained by going down the big one, that is analogous to lighting a bulb. There is, however, a problem with this analogy too. On a roller coaster, there are cars in only a few places, while in an electric circuit, the flow is continuous, like the flow of water.

Circuit Situations Revealed!

In Chapter 1, we presented some electric circuit "situations" and asked you to think about how they might work. The three situations were:

1. The Lamp, which will light when both its own switch and the switch on the power strip are ON;
2. The Yard Light, which will come on when either or both of the two switches are ON; and
3. The Stair Light, which can be turned either ON or OFF by either of two switches, located at the head and foot of the stairs, respectively.

In Chapter 1, we described these situations in Tables 1-2, 1-3, and 1-4, respectively, which show the condition of each light for every possible condition of the switches. In each case, there are two switches, which we can call "A" and "B". Using this shorthand, we have combined the data for all three cases into Table 2-3.

Table 2-3

THREE ELECTRIC CIRCUIT SITUATIONS

Switches		Lights		
A	B	Lamp	Yard Light	Stair Light
OFF	OFF	OFF	OFF	OFF
OFF	ON	OFF	ON	ON
ON	OFF	OFF	ON	ON
ON	ON	ON	ON	OFF

The Lamp Situation

In the Lamp Situation, both switches need to be ON for the lamp to work. Another way of saying this is that either of the switches, if OFF, could prevent the lamp from working. Suppose we have a complete circuit with no switches, such as the one shown in Figure 2-31. Now suppose that one of the wires is broken in two places. This would prevent the current from making a complete circuit, and keep the light off. Repairing only one of the breaks will not help, because the current will still be blocked by the other break. Only when both breaks are fixed will the lamp come on again. Now imagine that we use switches to replace the breaks in the wires. Using the language of schematic diagrams, this circuit would be shown as in Figure 2-39.

2-39: **The Lamp Situation**

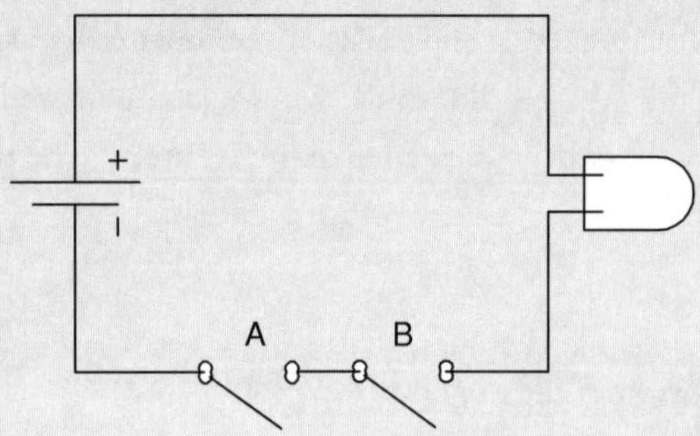

Let's see whether this circuit solves the lamp problem. Mentally put both switches in the ON (UP) position. There is now a direct path from the battery to the bulb, as in Figure 2-31. Move either switch (or both) to the OFF (DOWN) position, and the circuit is broken, just as in Figure 2-33, and no current flows to the bulb. This is the circuit we were looking for. In this case, the switches are said to be arranged *in series,* because all the current through one must go through the other.

The Yard Light Situation

2-40: **The Yard Light Situation**

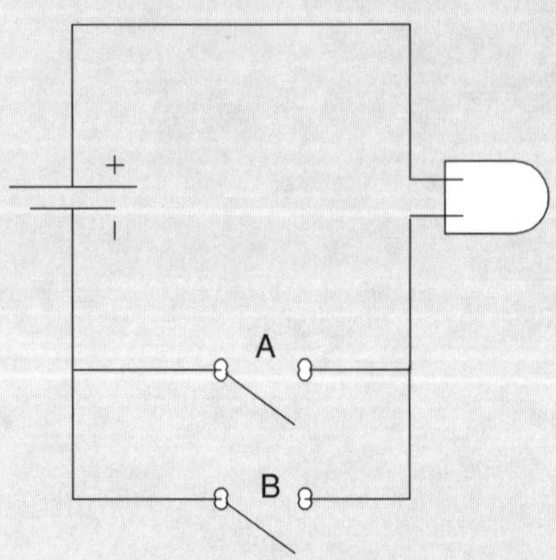

In the Yard Light Situation, either switch can make the bulb light. In other words, neither switch, by itself, can keep the light off. It's only when both switches are OFF that the bulb is shut down. How do you make a circuit that works this way? Suppose you were connecting a battery to a bulb with a long wire, but you were afraid that someone might accidentally break the wire. One strategy would be to run another wire along a different path, connecting the same battery and bulb. In case one of the wires got broken, the other would still be available to carry the current. For the bulb to turn off, both wires would have to be broken. Now replace the breaks in the two wires with switches. This would give you the circuit is shown in Figure 2-40.

Each of the two paths from the battery to the bulb will work as long as its switch is UP, i.e., ON. Because the two switches are connected end-to-end at both ends, they are said to be *in parallel.*

The Stair Light Situation

2-41: **Two-way switch installed**

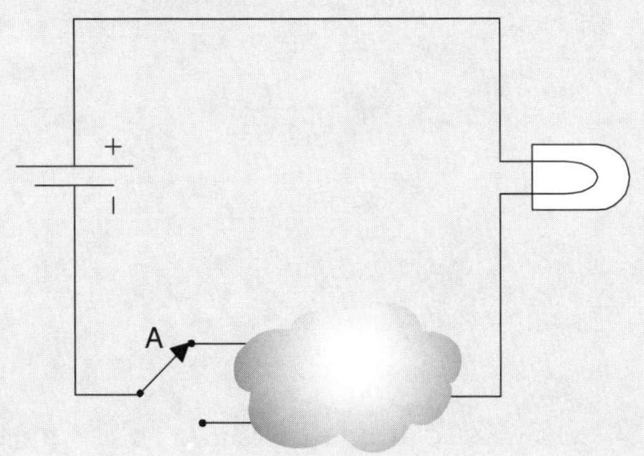

2-42: **Another two-way switch**

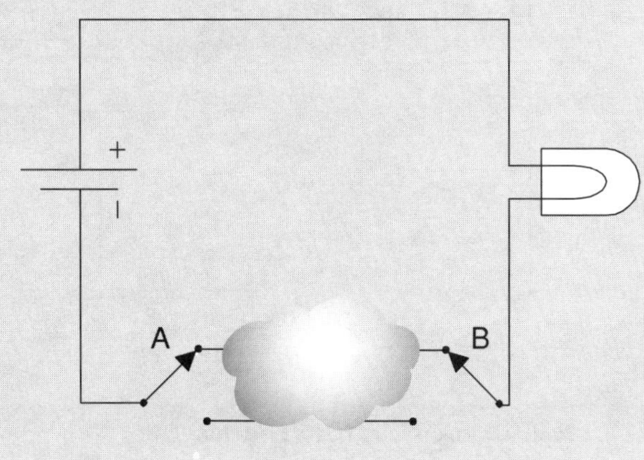

The Stair Light Situation is the most subtle of the three. There is no simple series or parallel connection that solves this one. In the Yard Light Situation, there were two ways to turn on the light: switch A or switch B. Each of these alternatives involves only one switch. If either one of the switches is ON, it doesn't matter what the other switch is doing.

In the Stair Light Situation, there are also two ways to turn on the light. But with the Stair Light Situation, each method depends on both switches. The two ways are:

- Switch A is OFF (DOWN) and switch B is ON (UP); or
- Switch A is ON (UP) and switch B is OFF (DOWN).

From this description, we can see that turning the lamp ON could involve either the UP or the DOWN position of each switch. This was not the case in the Lamp and Yard Light Situations, where only the UP positions were used to make the current flow. As a result, the Stair Light requires a different kind of switch, one that could carry current when either up or down. This "two-way" switch is shown in Figure 2-41. Note that the movable part of the switch, represented by an arrow, can attach to either of two terminals (UP or DOWN), each of which has a wire leading from it. We have put a "cloud" over these wires, because we haven't decided yet what to connect them to.

So far, we have installed one of the two switches, "A." Since both switches work exactly the same way, we will need another of these two-way switches for "B." As we have seen, current will need to go through both switches in order to light the bulb. So, our next step is to put a second two-way switch, "B," in series with "A." This additional switch is shown in Figure 2-42.
There is still a "cloud" between the two switches, because we haven't decided yet how to connect them.

Now, let's take just one of the ways the bulb could light: Switch A is DOWN, and switch B is UP. If we connect a wire from the DOWN terminal of switch A to the UP terminal of switch B, and put the switches in those positions, we should get a complete circuit, as shown in Figure 2-43. This is one of the two possible ways of making the bulb light.

2-43: **One way to turn on the Stair Light**

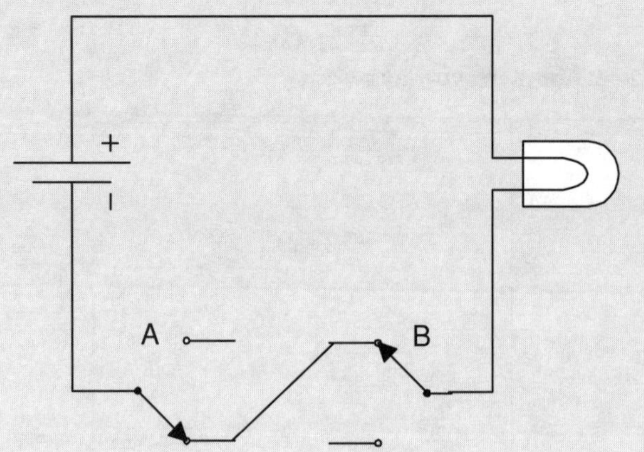

2-44: **Adding the other way to light the Stair Light**

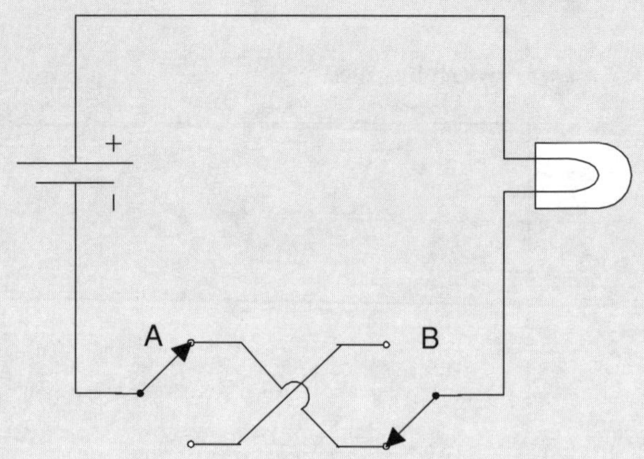

2-45: **Stair LIght circuit, both switched UP**

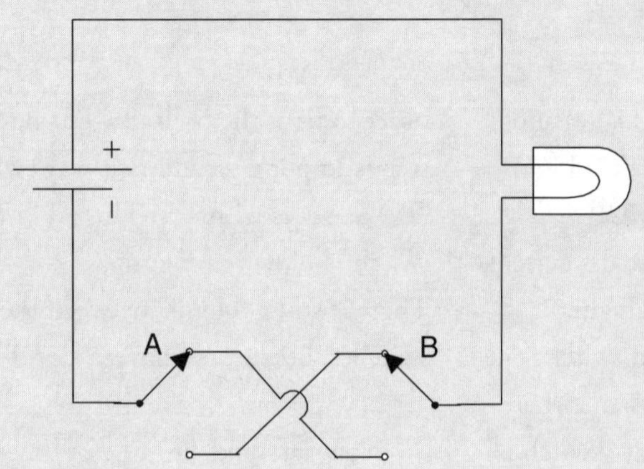

Next, we need to make the connection so that the bulb can light the other way: Switch A is UP and switch B is DOWN. This connection, which is simply the opposite of the one we just made, is shown in Figure 2-44. The little hook where the wires cross is a symbol that indicates "the wires are not touching each other."

Now let's check to see that this circuit works in all of its configurations. Figure 2-45 shows the case where both switches are UP. The bulb should not light, and it doesn't, because there is no complete circuit. The same would be true if both switches were DOWN. We have solved the mystery!

Beyond Electricity: Controlling Fluids and Mechanisms

Control and *control system* are two of the "big ideas" of technology. One way around the conceptual difficulties with circuits is to approach them using these broader concepts. From the point of view of controls, an electrical switch is just one of many devices that controls the flow of energy. A circuit is just one kind of control system, which processes energy based on information from its control device. We turn next to a discussion of controls, including both electrical and mechanical types.

In previous sections, we have looked at devices that control the flow of electricity. Many if not most of the controls we encounter daily are electrical, because electricity is very easy to move from one place to another and to transform into different forms of energy. However, even in our highly electrified world, there are other kinds of flows and motions that can be controlled. In this section we will consider other kinds of controls.

As we have seen, there are strong analogies between the flow of electricity and the flow of fluids, although there are also important differences. Although we normally think of fluids as liquids, gases are also fluids that flow in much the same way as liquids. If you have a gas stove, a gas furnace, a gas dryer, or a gas hot water heater, you can notice plumbing connections to these devices that look quite a bit like plumbing for water. The fluid in this case is natural gas. Compressed air is another gas that is frequently transported through pipes and hoses. Most service stations have compressed air hoses for filling tires.

Electrical switches are sometimes compared with shut-off valves, which are used in plumbing to start or stop the flow of gas or water. Faucets are a little different from shut-off valves because they have many positions between completely ON and completely OFF. With these devices, the user can adjust the flow from a trickle to a strong, steady flow, or anywhere in between. By analogy, with electronic devices, a shut-off valve is a *digital* device, like a switch, with only two positions, ON and OFF. A faucet is an *analog* device, which is continuously variable from ON to OFF. Other control devices for fluids include:

- a handle used to flush a toilet;
- a valve that permits air flow into an inflatable toy, raft, or air mattress;
- a burner control knob on a gas range or oven;
- an adjustable nozzle on a hose or shower head;
- a kitchen sink stopper/strainer (see Figure 2-46); and
- a nozzle that adjusts the spray from a hose or shower head.

Can you tell which of these are analog and which are digital?

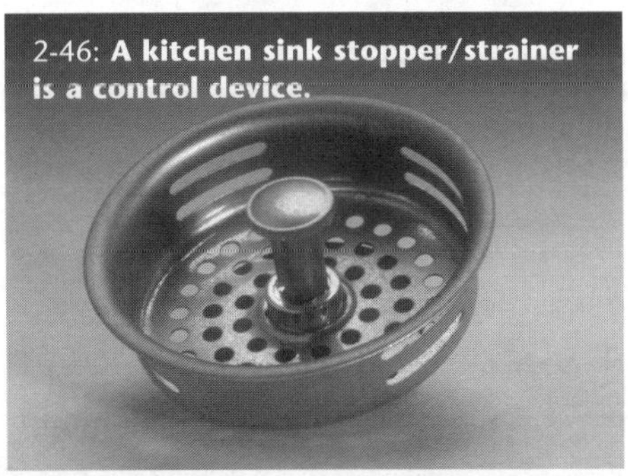
2-46: **A kitchen sink stopper/strainer is a control device.**

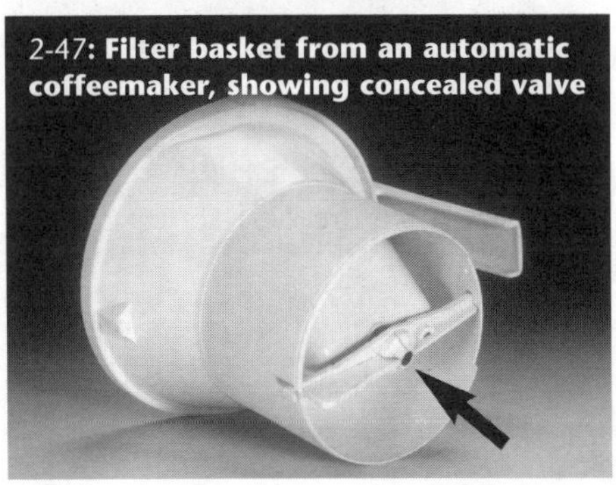
2-47: **Filter basket from an automatic coffeemaker, showing concealed valve**

2-48: **Depressing the pin opens a valve on the tire.**

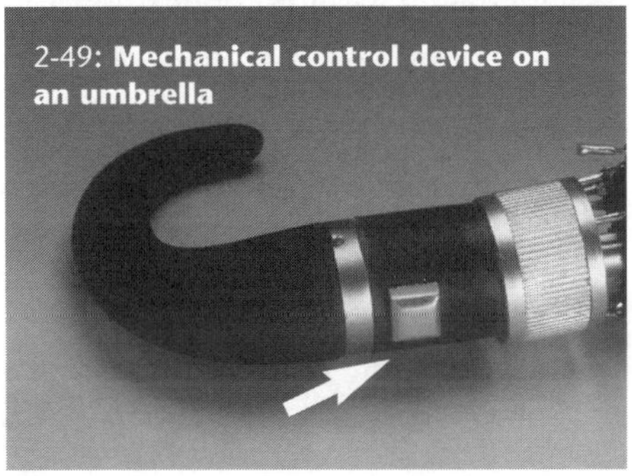
2-49: **Mechanical control device on an umbrella**

Some controls for fluids are of the hidden type. For example, Figure 2-47 shows a filter basket from an automatic coffee maker. The user puts the coffee grounds and filter into this basket and inserts it into a slot in the coffeemaker. Inserting the basket fully into the slot activates a little valve at the bottom center. The valve allows the water through the basket only when the basket has been properly positioned. Except that it operates on water, not electricity, this valve is very similar to the hidden pencil sharpener switch shown in Figure 1-55. A valve on a bicycle or automobile tire (Figure 2-48) is another example of a concealed control device for fluids. This type of valve has a little pin in it, which opens the valve when depressed by a matching pin at the end of the air hose.

We have also seen some controls in our discussion of mechanisms. A door lock (Figure 1-11) can be seen as a control that regulates the flow of people into and out of a room or house, much as a valve controls the flow of water or gas. An umbrella has a push-button (Figure 2-49) that releases a large spring, causing the entire umbrella to open. A bicycle handbrake (Figure 1-10) is part of a control system, operated by a lever on the handlebars, which causes the entire bicycle and rider to come to a stop.

The umbrella release button, cylinder lock, and bicycle handbrake are all examples of mechanical control devices. In each case, a small amount of energy at the input (the control) unleashes a much greater flow of energy elsewhere. A graphic example of a mechanical control device is the mousetrap, shown in Figure 2-50. If the trap is set properly, the unwitting "user" (a mouse) exerts ever so little energy to release the trap, and WHAM!!! A much larger burst of energy is released by the spring!

A similar description applies to electrical switches and fluid controls. A tiny bit of energy at the input—the flick of a switch or turning of a valve—can control a large flow of electrical or mechanical energy elsewhere. Pushing an elevator button, which is nearly effortless, can cause a huge motor to lift thousands of pounds. Similarly, depressing an accelerator pedal slightly forces more gas and air into the engine, causing an entire car to gain speed. The elevator button is an example of an electrical control, while the accelerator pedal is a fluid control.

Some mechanisms would not be considered examples of controls because the input and output use about the same amounts of energy. In these devices, there is a direct link between the input and output. The wastebasket, eyelash curler, ice cream scoop, egg topper, pizza tray holder, etc., are not really control devices for this reason. In these cases, the control is not really separate from the thing controlled. A control system has to have some way of storing energy, so that the control device can release it on demand. In a mousetrap or umbrella, mechanical energy is stored in a spring. A battery performs the same function in an electric circuit, and a tank stores gravitational energy for a toilet by holding water above the level of the bowl.

2-50: **Triggering the control device on a mousetrap releases a burst of energy.**

Controls Defined

We have given some examples of what we call controls, but we have not explained what the term really means. Because the word *control* has so many meanings in everyday life, it can be difficult to pin down what is and is not a control. In technology, *control* has a precise meaning. It involves two activities: the flow of information and the flow of energy.

Imagine a professional wrestler who has never learned the script for the bout. A prompter, standing by the side of the ring, whispers each instruction: "Grab him by the waist... Lift him up... Twirl him around your head...." The prompter is sending information, in the form of commands, but is himself expending very little energy. The wrestler, on the other hand, is processing a great deal of energy, but only on the basis of information received from the prompter. One member of the team, the prompter, has plenty of information, and decides what should be done at any moment, but lacks the power to do it. The other member, the wrestler, lacks any information of his own, but has the strength to carry out the commands. The prompter is playing the role of the control device, while the wrestler handles the energy flow that is being controlled.

A control system, by this definition, has a control input, which uses little energy to communicate information. It also has an energy processor, whose work is dictated by the control information. Figure 2-51 shows this relationship.

In any control system, we should be able to identify three basic elements:

- The control input;
- The energy processor that is controlled; and
- The form of the energy that is processed.

Table 2-4 gives some examples of common control systems, with each of these elements identified.

2-51: **Control system block diagram**

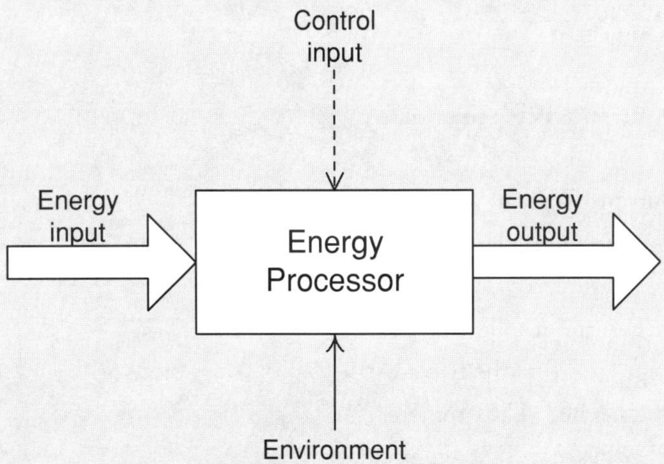

In all but one of the control systems we have examined so far, a human user supplies the information to the control input. (The one exception is the mousetrap, where the unfortunate "user" is a rodent.) Consequently, these are sometimes described as *manual control systems.* There are many important control systems that do not depend on human inputs, because they generate their own information. These are known as *automatic control systems.*

Table 2-4

SOME CONTROL SYSTEMS

System	Control input	What is controlled?	Nature of energy flow
Mousetrap	Release lever (activated by mouse)	Trap arm	Release of energy stored in spring
Stovetop	Burner control knob	Gas burner	Flow and combustion of gas, releasing heat energy
Car	Accelerator pedal	Engine	Flow and combustion of fuel; energy of motion of the car
Door lock	Key	Door	Energy of motion of people passing through
Electric circuit	Switch	Light bulb	Electric energy supplied by battery; light energy radiated by bulb
Hose	Nozzle	Water spray	Energy of flowing water

Automatic Controls

To see the difference between an automatic and a manual control system, let's revisit our professional wrestler and his prompter. Suppose the wrestler's opponent is trying his best to win the match without following any prepared script, and that the two are about evenly matched. The promoter of the event has instructed the prompter of the first wrestler to prolong the match as long as possible, for commercial reasons. What strategy should the prompter follow in order to prevent his man from either winning or losing?

Recall that in the original example, the prompter made his wrestler follow a predetermined script. In this new case, the opponent adds an unpredictable element to the situation. If the prompter's man is winning, the prompter should not let him continue beating up his competitor and risk ending the bout prematurely. On the other hand, if the same wrestler is losing, the prompter should give him the moves he needs to get back in the match.

In this situation, where an adversary provides a varying context, the prompter sizes up the situation before issuing instructions. He always tries to reverse whatever is happening at the moment in order to keep the bout on a neutral course. If his man is winning, he tells him to slow it down, and vice versa. Translating this example to the realm of devices, imagine a system where the human prompter is replaced by a circuit and/or a mechanism. Such a device would use information about the state of things to turn energy flows on or off. This type of setup is called an automatic control system, because it operates on its own, basing its actions on its own data about each current situation.

2-52: **Automatic control system**

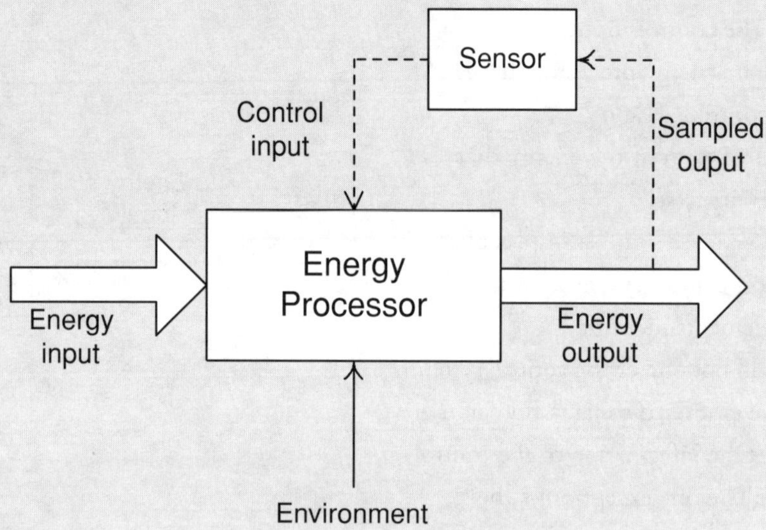

The best known example of an automatic control system involves the thermostat system used to control heating and air conditioning in a house. Thermostats are also used to maintain the desired temperature in refrigerators and in some ovens, irons, and electric blankets. After the user sets the desired temperature, the thermostat monitors the actual temperature continuously. The monitoring device, in this case a thermometer, is called a *sensor.* Once the temperature reaches the desired level, the thermostat turns the heating or cooling device off.

Every automatic control system relies on some sort of sensor to provide information about the outcome. This information feeds back to the control device, which in turn modifies the energy flow accordingly. For this reason, automatic control systems are sometimes called feedback systems. A block diagram of an automatic control system is shown in Figure 2-52. The dashed line from "Sampled output" to "Control input" is sometimes called the *feedback loop.* It transfers information only. Because of the presence of this loop, which returns back from the output to the input, a feedback system is sometimes called a *closed loop system.* The manual control system, which lacks feedback, is called an *open loop system.*

Let's look at how the information from the thermometer is used in a thermostat-controlled home heating/cooling system. If the air is too warm, the thermostat senses this and tells the air conditioner to come on; conversely, if it the temperature is too low, it activates the furnace. In either case, the actual conditions are the basis for deciding what to do next. The purpose is always to restore the temperature to a comfortable setting. The thermostat plays the role of the prompter—it controls the energy sources that bring the temperature back to the desired value. The energy sources are the furnace and the air conditioner, and these are analogous to the wrestler who obeys the instructions of the prompter.

Who plays the role of the adversary in the home heating/cooling system? Automatic control systems are designed to provide shelter from unpredictable changes in the natural and artificial environments. In the case of a thermostat system, the "adversary" includes both natural events, such as changes in outdoor temperature and sunlight; and human activities, such as opening and closing windows and doors, and turning lamps and appliances on or off. All of these affect the indoor temperature. These factors that make the control system necessary can be collected under the heading *environment.*

Manual and the automatic control systems both respond to changes in the environment, but they do so differently. A manual control system depends on a human user to recognize that the environment has changed and to compensate using the control input. For example, if a car is slowing down because it is going uphill, the driver recognizes this fact and operates the accelerator pedal. An automatic system senses changes in the environment and operates the control input automatically, eliminating the need for a human operator in the feedback loop. For example, a car with cruise control does not depend on the driver to use the accelerator. Instead, it uses its own data about the speed of the car to speed the engine up or slow it down.

Table 2-5 gives some examples of automatic control systems found in everyday life.

Table 2-5

EXAMPLES OF AUTOMATIC CONTROL SYSTEMS

System	Control input	Sensor	Energy flow that is controlled	Environment against which system operates	Goal of system
Home temperature control (thermostat)	ON/OFF switches of furnace and air conditioner	Thermometer	Furnace and air conditioner	Natural and artificial variations in temperature	Maintain indoor temperature at desired value
Automotive cruise control	Throttle opening	Speedometer	Car engine	Friction, variations in road surface, hills	Maintain desired speed
Tank toilet	Valve, which allows water to fill in tank	Float and arm, attached to valve	Flow of water into tank	Flushing of toilet, evaporation, leakage	Maintain water level in tank
Automatic sprinkler for fire protection	Glass bulb, blocking water supply	Liquid in bulb expands when hot, breaking bulb	Flow of water into room	Possibility of fire	Extinguish fire from room
Steam iron with "automatic shutoff"	Automatic shutoff switch	Motion sensor, detects that iron has not been moved for 10 minutes	Heat flow through iron	User forgets to turn it off	Prevent iron from starting fire by heating one place too long
Exposure control of automatic camera	Lens opening (f-stop)	Light sensor	Amount of light allowed through aperture and lens	Amount of ambient light varies in time and space	Prevent film from being over- or under-exposed

Feedback Control in Nature and Society

All of the examples in Table 2-5 are from the world of technology. Nature, too, abounds in feedback systems. Here are some simple experiments you can do to explore how feedback works in your body:

1. **Try to touch your toes, standing with your back to a wall.** You can't do it! Why not? To answer this, watch from the side while someone whose back is not against a wall touches his or her toes. Notice how the rear end moves back as the person leans down. This movement is necessary to keep one's balance. Otherwise, too much of the weight is leaning forward, and it is impossible to avoid falling forward, which is what happened when your back was against the wall. When you touch your toes normally, feedback tells your muscles to move your buttocks back, so you won't fall over.
2. **Stand on one foot.** With a little effort, you can probably keep your balance with one foot off of the floor. (See Figure 2-53.) However, you may notice a series of jerky movements, as you lean too far in one direction, then another, and your feedback system compensates by telling your muscles to move in the opposite direction. Now, close your eyes. It is now much harder to remain standing on one foot, and you will probably have to use your arms to keep your balance. (See Figure 2-54.) That's because your major source of feedback information—vision—is no longer active.
3. **Trace a drawing using a mirror.** On a piece of paper, draw a simple figure such as a square or star. Then place the drawing in front of a mirror. Looking only in the mirror, and not at the drawing directly, try to trace the figure you have drawn, as shown in Figure 2-55. It is much harder than normal tracing because the mirror image has disrupted the normal feedback from eye to hand.

Here are some more examples of feedback in the human body:

- Body temperature is regulated by a complex system that tries to maintain it at 98.6 degrees, regardless of outside temperature or level of physical activity. This system causes sweating to occur when the body temperature starts to rise. It also causes shivering when the temperature drops too low.
- The pulse and breathing (respiration) rates are controlled by another complex feedback system whose goal is to supply sufficient nutrients and oxygen to all parts of the body. The system does this by increasing the

2-53: **Keeping one's balance on one foot**

2-54: **Keeping one's balance on one foot is harder with eyes closed.**

2-55: **Trying to trace a drawing reflected in a mirror**

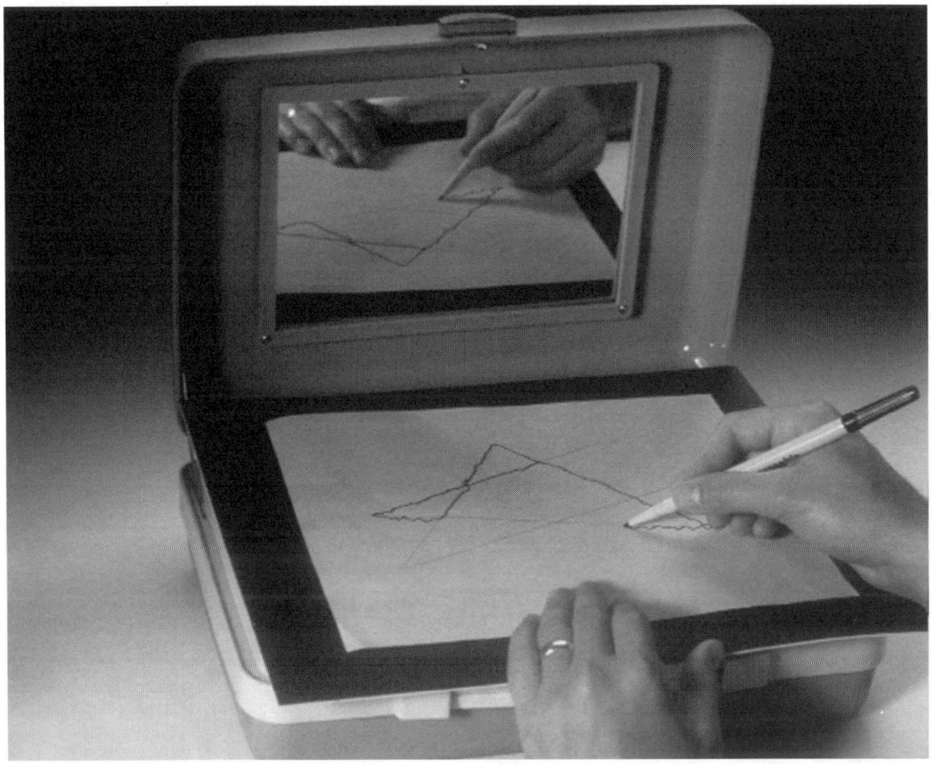

pulse and respiration rates during periods of intense activity and decreasing them when the body is at rest.

- The light input to the eye is controlled by the pupil. It plays exactly the same role as the aperture (or "F Stop") in a camera. In broad sunlight, the pupils close, so that little of the light will enter. In a darkened room the pupils open up to let in as much light as possible. The adjustment can take some time, and the system can be overwhelmed by rapid changes in the environment. For example, the light suddenly bothers you when you emerge from a tunnel or building on a bright, sunny day.

The human body is only one example of a natural system. Every biological organism utilizes many different forms of feedback control to maintain its internal environment against changes and challenges from outside. A plant that grows towards the light is regulated by a complex feedback system that senses light and uses this information to control growth.

Feedback is also an essential element in longer-term and larger-scale natural systems such as biological evolution, plant and animal populations, ecosystems and the global climate system. Let's look briefly at this last system.

The earth's atmosphere contains a delicate balance of gases, the most important of which are oxygen and carbon dioxide. Green plants have produced nearly all of the oxygen, while most of the carbon dioxide is released by the ocean or by animal life. For millions of years, the amounts of these gases have been prevented from changing rapidly by complex feedback mechanisms. As carbon dioxide concentrations increase, more plant growth is stimulated, which in turn removes some of the carbon dioxide from the atmosphere. On the other hand, increases in oxygen benefit animals, which replace the excess oxygen with carbon dioxide. Recently, however, the destruction of forests and burning of fossil fuels threaten to overwhelm this natural feedback, leading to sharp increases in carbon dioxide and global warming.

Feedback is also a major factor in nearly every social enterprise, from the very smallest to the very largest. A teacher routinely uses feedback in the design of lessons, of student groups, of classroom arrangements, etc. She tailors each of these to her students' needs by continuously monitoring what works and what doesn't, and making corrections based on this information. What can make lectures so stultifying is the lack of feedback from the students back to the teacher. The same is true of instruction driven by standardized tests, because it takes so little account of students' needs and interests. Larger social systems, such as cities and nations, must also depend on feedback from ordinary people in order to be successful in addressing their needs. Social systems that do not use feedback can legitimately be described as dictatorships.

Popular Conceptions of Feedback Systems: Do You Know How Your Thermostat Works?

Although feedback and automatic control are key concepts in modern science and technology, they are probably not well understood by most people. Although words like feedback and loop have crept into everyday language, their technical meaning is often lost. "Can I have your *feedback?"* (Translation: "Are there any comments?") "These days, I am completely out of the *loop!"* (Translation:" I am not in the inner circle.")

So, what do people really know about feedback and control? Kempton (1987) did a study on "folk theories" of home heating control. Based on interviews and actual recordings from homeowners thermostats, he describes two different theories people have about these devices:

1. The "valve theory" assumes that a thermostat works like a faucet or accelerator pedal. The higher you set it, the more rapidly the heat gets pushed into the room.
2. The "on-off theory" says that the thermostat turns the furnace on whenever the temperature in the room is below the thermostat setting. Turning the setting up makes the furnace operate longer, because it will take more time to reach the desired temperature, but not faster, because the furnace can only operate at one speed.

The "on-off" theory gives the most accurate description of how a thermostat really works, but the valve theory is the most widely held. Most people think that they can make the room warm up faster by turning the temperature setting way up.

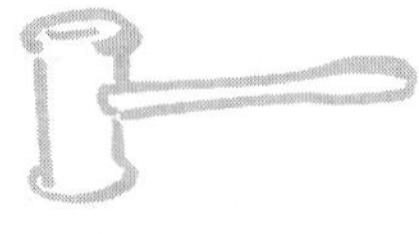

Chapter 3

Activities

The activities in this chapter are designed to give students direct experience with mechanisms and circuits. The activities were created and tested by classroom teachers. Many of their experiences with these or similar activities are described in Chapter 4, "Stories."

Activities 1-9 deal with mechanisms and the concepts related to mechanisms. Activities 10-14 deal with circuits. The activities are designed to give students experience with many of the concepts discussed in Chapter 2, "Concepts."

All of the activities are correlated to standards in Science, Mathematics, and English Language Arts. The standards are listed by number with each activity; the standards themselves are listed at the end of the chapter.

ACTIVITIES AT A GLANCE

Level	Activity Title	Page	What Students Learn About Mechanisms and Circuits		
			What They Are	How They Work	Redesign/Design
Introductory	What Is a Mechanism?	74	×		
	Be a Mechanism Detective			×	
	Mechanism Scavenger Hunt	76	×		
	What Does a Tool Do?	77	×		
	Can You Guess My Categories?	79			×
Intermediate	Ins and Outs of Inputs and Outputs	80		×	
	How Do Levers Make Work Easier?	82		×	
Advanced	Simple Machines	85	×		
	How Does a Retractable Ballpoint Pen Work?	89		×	
	Make a Model of a Mechanism	91		×	×
	Conductors and Insulators	94	×	×	
	Electric Switches	96	×	×	×
	Two Switches, One Lamp	98		×	×
	Electric Circuit Board Game	100		×	×
	Water-Level Alarm	102		×	×

Activity №1

What Is a Mechanism?

Grade Level
K – 5

Prerequisites
None

Overview
This is an introductory activity that uses discussion and concept-mapping to have students articulate what they know and believe about mechanisms.

Concepts
- Mechanisms are found in everyday life.
- All mechanisms have moving parts.

Vocabulary
Mechanism

Skills
- Brainstorming
- Accessing prior knowledge

Standards
- Standards for the English Language Arts:12
- Benchmarks for Science Literacy: 1B
- National Science Education Standards: A

Time Needed
45 minutes

Materials
- Chart tablet
- Markers

Procedure
1. Bring students together for a group discussion.
2. Use questions like those below to elicit what students know and believe about mechanisms and to help them discover the idea that all mechanisms have moving parts. Guide students to describe places where mechanisms are used (e.g., kitchen, bathroom), examples of their functions (cut, slice), and kinds of motion (push, pull, turn).
   - What do you think the word mechanism means?
   - What do mechanisms do? What do we use them for?
   - What are some examples of mechanisms in our classroom?
   - Where else can you find mechanisms?
3. Record their responses on a word web (see page 75) and/or a K-W-L chart ("What do we **K**now?", "What do we **W**ant to know?", "What did we **L**earn?").
4. During this discussion, do not label students' answers as right or wrong. Instead, return to your record of this discussion as your study of mechanisms progresses. Then you can give students the chance to revise their own ideas and answers as they gain more knowledge and understanding.

Tips
Help students identify the characteristics of mechanisms by showing them examples (e.g., a stapler or manual pencil sharpener) and asking them to describe what they see. Ask how the device is used, what happens when it is being used, which parts move, and how they move.

Sample Word Web for Activity #1

"What Is a Mechanism?"

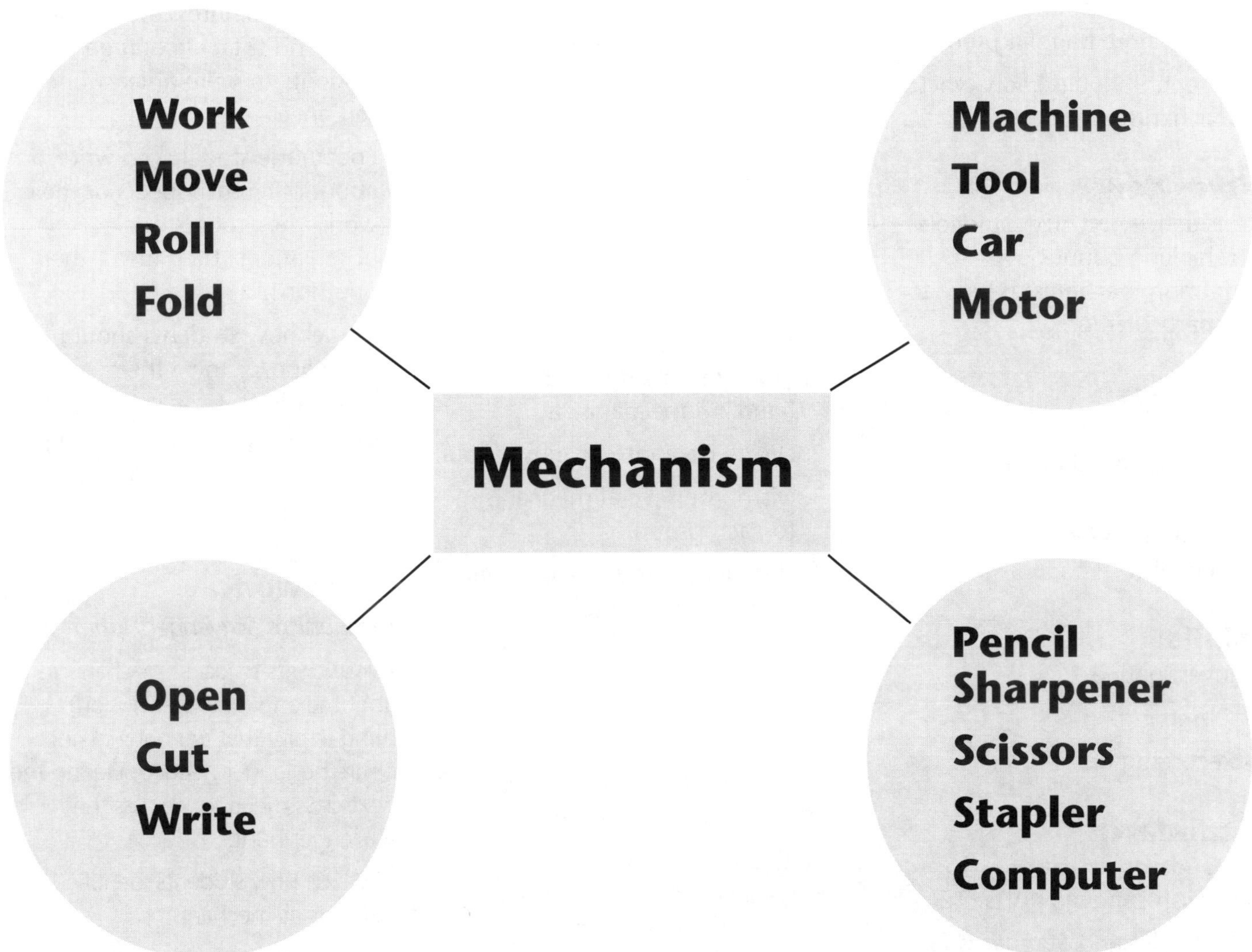

Activity No 2

Be a Mechanism Detective

Grade Level
K-5

Prerequisites
- Students must know safety rules before handling tools.
- Students should know what mechanisms are.

Overview
This activity involves students in thinking about how common mechanisms work to do particular jobs.

Concept
Common mechanisms work based on cause-and effect-relationships.

Vocabulary
Mechanism

Skills
- Investigating
- Observing
- Recording data

Standards
- Standards for the English Language Arts: 12
- Principles and Standards for School Mathematics: G4
- Benchmarks for Science Literacy: 1B, 2A, 2C
- National Science Education Standards: A

Time Needed
45 minutes

Materials
- Common mechanisms such as hole punchers, nutcrackers, ice tongs, garlic presses, scissors (for grades 4 and 5, more complex mechanisms can be used)
- Chart tablet
- Markers, crayons

Procedure
1. Review safety rules for handling mechanisms.
2. Divide class into groups of four.
3. Distribute several mechanisms to each group.
4. Have each group explore and examine their objects. Prompt them with questions such as "What does it do?" "What do you have to do to it to make it work?"
5. Have students trace or draw and describe their objects in writing.
6. Depending on grade level, have groups record or keep notes on their group discussions.
7. Have individuals and groups share their results.

Tips
- For one teacher's experience with this activity, see Chapter 4 ("Stories"), page 118.
- Allow 30 minutes for investigation, and then encourage students to write at least one sentence.
- Encourage students to write about different aspects of their work: the process, their observations, their ideas, their questions.
- Model how students should draw and/or trace their mechanisms.
- Record safety rules on an experience chart for future reference.

Homework
Mechanism Scavenger Hunt

Ask students to locate mechanisms at home and (with permission) bring in any that are not working or are no longer needed. During the next class meeting, discuss the scavenger hunt.

- Discuss how students identified objects as mechanisms.
- If students are unsure about whether something is a mechanism, encourage the class to discuss it and come to a consensus.

Activity No 3

What Does a Tool Do?

Grade Level
K-5

Prerequisites
Students should be aware of safety issues before using tools.

Overview
This is an introductory activity for the early grades or a getting started activity for the middle and upper grades. Children try to define the word "tool" and trace the positions of a simple tool in its open and closed positions.

Concept
Mechanisms can be found in tools.

Vocabulary
- Mechanism
- Tool
- Open
- Closed
- Rest
- Operating

Skills
- Investigating
- Observing
- Recording data

Standards
- Standards for the English Language Arts: 12
- Principles and Standards for School Mathematics: G1
- Benchmarks for Science Literacy: 1B, 2A, 3A
- National Science Education Standards: A, B

Time Needed
45 minutes

Materials
- Worksheet #3: "What Does a Tool Do," one for each student
- Chart tablet
- Markers
- Hole punchers (for grades K-2), one for each group of 3 to 5 students
- Vise grips (for grades 3-5), one for each group of 3 to 5 students

Procedure
1. Review the class's discussions about mechanisms in connection with Activities 1 and 2, focusing on what students discovered about what a mechanism is and where mechanisms can be found.
2. Ask, "What is a tool?"
3. Record all student responses on an experience chart.
4. Place students in groups of 3 to 5.
5. Give out hole punchers (Grades K-2) or vise grips (grades 3-5). Elicit responses about the object by asking such questions as,
   - Is this a mechanism?
   - Is it a tool?
   - What do you think it is used for?
6. Record students' responses on the experience chart, without labeling them as right or wrong.
7. Distribute worksheets, one to each student.
8. Have students trace the mechanism in the open (rest) position and in the closed (contracted) position.
9. Encourage groups to discuss how it might work and its possible uses. Depending on grade level, groups can record their ideas.
10. Bring all groups together to share their experience and ideas.
11. Record students' responses and compare them with their responses in the earlier discussion. Revise the experience chart to reflect students' new insights and understanding.

Tips
- For one teacher's experience with this activity, see Chapter 4, page 125.
- For younger students, model the tracing process in the open and closed positions. Label all positions.
- For ESL students, use vocabulary ("open," "closed," etc.) in both languages.

Worksheet for Activity #3

"What Does a Tool Do?"

Name ______________________________

Draw your tool when it is open and when it is closed.

Open position	**Closed position**

This tool is called ______________________________

What does this tool do? ______________________________

____

____

____

How does the mechanism work? (Write it here or on your pictures.) ______________________________

____

____

____

Activity No 4

Can You Guess My Categories?

(Who Am I and What Do I Do?)

Grade Level
K-4

Prerequisite
Knowledge of a variety of ways to classify

Overview
This is a simple sorting activity made into a game using a variety of mechanisms. Students must sort mechanisms, try to name each one, and determine what it is used for.

Concept
Mechanisms can be sorted and categorized according to their functions as well as other criteria.

Vocabulary
- Category
- Categorize
- Sort

Skills
- Collecting data
- Analyzing data
- Organizing data
- Sorting and classifying
- Recording data

Standards
- Standards for the English Language Arts: 7, 12
- Principles and Standards for School Mathematics: A1
- Benchmarks for Science Literacy: 1B, 1C, 2C
- National Science Education Standards: A

Time Needed
2 sessions, 45 minute each

Materials
- Office and household tools or gadgets (e.g., nail clippers, lipstick, gluestick, eyelash curler, pencil sharpener, garlic press, etc.)
- Paper
- Pencils
- Notebooks
- References on simple machines or reference list of mechanisms
- Index cards
- Experience chart

Procedure
1. With the whole group, discuss what it means to sort things by "category." If necessary, model categorizing a group of objects or people.
2. List responses on chart paper as students share ideas.
3. Divide students into groups.
4. Give each group a tray with several mechanisms for the students to classify.
5. Ask students in their groups to name each mechanism and discuss what it is used for.
6. If the students want to know what an unfamiliar object is, hold it before the class. If no one can name it or tell how it is used, ask them how they could find out what it is. Demonstrate the mechanism's use, if possible. If students are still unable to figure out what the mechanism is used for, record their questions on chart paper and return to the object later for further research.
7. After everyone in each group has examined the objects, the group decides how they should be categorized.
8. Students write the name of each category on an index card. (For younger students, have students tell you the category names so you can write them on cards.)
9. Ask the groups to present the objects in each category to the class without naming the category.
10. Invite the other groups to guess what the objects in each category have in common.
11. After each group has presented its grouped objects, discuss other possible categories, focusing on what the objects have in common.
12. Record all questions that may arise. Have references available on simple machines. Encourage students to record their experience in their journals.

Tips
- If categorizing becomes difficult, students should practice by sorting various objects available in the classroom—e.g., crayons, toys, books, etc. They can also sort students by what they have in common—e.g., light or dark hair, kind of shoes, glasses or no glasses, etc.
- Encourage students to categorize their mechanisms by what they do or how they work rather than superficial characteristics such as size or color.

Activity No 5

Ins and Outs of Inputs and Outputs

Grade Level
3-6

Prerequisite
Understanding of categorizing mechanisms and the ways mechanisms can be operated

Overview
In this activity, students examine common mechanisms to find their inputs and outputs, and to identify the motions executed at both input and output.

Concept
Mechanisms operate based on input and output.

Vocabulary
- Input
- Output

Note: *For an introduction to these terms and concepts, see the Glossary at the back of this Guide.*

Skills
- Problem solving
- Labeling

Standards
- Standards for the English Language Arts: 12
- Principles and Standards for School Mathematics: G1
- Benchmarks for Science Literacy: 1B, 3A, 3B, 11A
- National Science Education Standards: A, B

Time Needed
90 minutes

Materials
- Worksheet #5, "Ins and Outs of Inputs and Outputs"
- Common mechanisms such as a can opener, bottle opener, eyelash curler, ice cream scoop, wrench, nail clippers, tweezers, staple remover, etc.
- Construction paper
- Pencils and markers

Procedure
1. Choose one of the mechanisms from the classroom collection and demonstrate its operation in front of the class.
2. Lead a discussion about how the device works and what it does. Ask such questions as, "What makes this mechanism work?" Use the terms *input* and *output* to describe the operation of the mechanism.
3. Record students' responses and questions on an experience chart.
4. Draw the mechanism you've demonstrated on the experience chart. As you and your students describe and discuss the mechanism, label the drawing using the terms *input* and *output*.
5. Divide students into small groups and have each group pick an object to study. Ask each group to locate both the input and the output. If students are still unclear about what these terms mean, use prompts such as:
   - "What is the part that I use to make it work?"
   - "What is the part that actually does the job that I want this mechanism to do?"
6. Ask students to work together in their groups to draw diagrams of their objects, labeling the input and the output. Have them trace the motions of both input and output as the device is operated, and describe the differences between their motions.
7. Ask each group to share what it has found with the class.

Tips
- Use simple objects so it's easier to identify the input/output relationship.
- While students are sharing their drawings, encourage them to use the appropriate vocabulary: *input* and *output.*
- If an overhead projector is available, students can demonstrate their devices by placing them directly on the projector. The object will be seen on the screen in silhouette.
- Encourage students to write about their experience with this activity in their journals, including what was easy and hard, what they learned, and what they don't understand.
- Always remember to review safety rules.

Worksheet for Activity #5

"Ins and Outs of Inputs and Outputs"

Name ______________________________

Draw and label your mechanism using the terms *Input* and *Output*.

What is the name of this mechanism? ______________________________

What part of the mechanism do I have to move to make the mechanism work? ______________________________

What part of the mechanism does the job? ______________________________

Activity № 6

How Do Levers Make Work Easier?

Grade Level
2-6

Prerequisite
General understanding of the terms *input* and *output* as they relate to how a mechanism works.

Overview
This activity teaches the law of the lever, using a meter stick, a pencil, and a book.

Concepts
- A lever is a simple mechanism that can be used to assist in lifting heavy objects.
- A lever requires a fulcrum.
- The effort required to lift a load is related to the position of the fulcrum in relation to the load.

Vocabulary
- Input
- Output
- Effort
- Load
- Fulcrum
- Lever

Skills
- Predicting
- Collecting data
- Analyzing data
- Organizing data
- Communicating ideas and information

Standards
- Standards for the English Language Arts: 12
- Principles and Standards for School Mathematics: A1, DA & P1, DA & P3, C3, M1
- Benchmarks for Science Literacy: 1B, 2A, 3A
- National Science Education Standards: A, B

Time Needed
45 minutes

Materials
- Worksheet #6, "The Book-Lift Challenge," at least one for each student
- Meter sticks or yard sticks (one for each group of four students)
- Pencils (one for each group of four students)
- Hardcover books (one for each group of four students)
- Notebooks
- Chart paper

Procedure
1. Place a hardcover book on a desk so that every one can see. Then show students a meter stick and a pencil.
2. Tell students that they are going to work in groups. Each group will have a book, a meter stick, and a pencil to solve a challenge. Write this challenge on the board: "Use the meter stick and the pencil to lift the book off of the desk or floor. The pencil must stay on the desk or floor at all times. Some part of the meter stick must always rest on the pencil. The end of the meter stick you push on cannot be the end that lifts the book."
3. Divide class into groups of three or four.
4. Distribute one book, one pencil, and one meter stick to each group. (Each group will need floor or desk space for their experiments.)
5. Distribute multiple copies of Worksheet #6 to each group. Explain to students that they should use the worksheets to describe all the different ways they tried to lift the book. Have extra worksheets available.

6. Give students about 10 minutes to work on the challenge. Help groups that are having trouble collaborating or conceptualizing solutions. If any groups don't seem to be coming up with a solution that involves creating a lever, ask questions to guide them toward that solution.
7. After about 10 minutes, ask the groups one at a time to describe what they did. During this discussion, use the term *input* to describe the pressure that was applied to the ruler and *output* to describe the lifting of the book.
8. Review the meanings of those terms from previous activities. Then explain that what students have made is a *lever*. Identify the pencil as the *fulcrum* of the lever. Finally, explain that with a lever, the force at one end is called the *effort* and the weight at the other end is called the *load*.
9. Set up the lever in front of the class, with one end of the meter stick under a book. Ask students if it makes any difference where on the meter stick you push down in order to lift the book. Is it easier or harder when you push close to the pencil than if you push at the end of the meter stick? Tell students their next challenge is to answer that question.
10. Give students another 10 minutes to investigate this question. Remind them to keep track of what they do on worksheets.
11. After about 10 minutes, bring the groups together again and discuss the results of their investigations. What they will have discovered is the *law of the lever* and the principle of *mechanical advantage.* Record questions, responses, and observations on chart paper for further investigation and reference.

Extensions

- Ask students to explore the effect of moving the fulcrum closer to and further from the object being lifted.
- Ask students to think of ways a lever could be useful in everyday life.
- Ask students to explain how various tools use levers to make work easier. Some examples are pliers, scissors, can opener, nail clipper.

Tips

- Introduce new vocabulary (lever, fulcrum, load, effort) in context as you demonstrate the lever and as you discuss students' problem-solving processes with them.
- Relate the lever to other mechanisms students have examined. Point out that the fulcrum is the part of the lever that does not move when the lever is being used.

Worksheet for Activity #6

"The Book-Lift Challenge"

Name ______________________________

How did you use the pencil and the meter stick to lift the book? (Describe it in words.)

____

____

____

Draw a picture of what you did. Show where you placed the book, the pencil, and the meter stick. Show what you did (the input) to lift the book (the output). Label the input and the output on your picture.

Activity № 7

Simple Machines

Grade Level
3-6

Prerequisites
Some experience with the study of simple machines

Overview
This activity connects the traditional science topic of simple machines with the study of mechanisms. It introduces the six simple machines: lever, wheel and axle, wedge, pulley, inclined plane, and screw. (All mechanical labor-saving devices are variations of those six.)

Concept
Simple machines make everyday tasks easier

Vocabulary
- Simple machine
- Lever
- Wheel-and-axle
- Wedge
- Pulley
- Inclined plane
- Screw

Note: *For an introduction to these terms and concepts, see Chapter 2 ("Concepts"), page 43, as well as the Glossary at the back of this Guide.*

Skills
- Understanding relationships of parts to whole
- Communicating ideas and information
- Collecting, analyzing and organizing data
- Observing
- Recording data

Standards
- Standards for the English Language Arts: 12
- Benchmarks for Science Literacy: 1B, 2A
- National Science Education Standards: A, B, E

Time Needed
90 minutes

Materials
- Worksheet #7: "Simple Machines"
- Index cards
- Chart tablet
- Collection of objects that incorporate simple machines (e.g., door stop, wheel-and-axle, pulley, screw, lever) as well as more complex mechanisms (e.g., spool, toy car, umbrella, folding chair, tape player, typewriter, doll carriage)

Procedure

1. If necessary, review your previous work with simple machines.
2. Have students sit in a circle. Place a variety of devices in front of them including some that are simple machines (e.g., wheel, screw, inclined plane, lever) and some that are not (e.g., toy car, umbrella)
3. Write the following vocabulary words on the chart tablet as well as index cards (one per card): "pulley," "inclined plane," "lever," "screw," "wedge," "wheel," "axle."
4. Place the index cards face up in a row inside the circle of students. Ask individual students to place an object under the corresponding index card. For example: doorstops would go under the index card labeled "inclined plane."
5. Ask which of the objects are simple machines.
6. Write students' responses on the chart tablet or chalkboard.
7. Divide students into small groups. Ask each group to pick one simple machine and devise a use for it, showing how it could help to make a job easier. For example, they could suggest moving a heavy desk using a lever, or using a ramp to move a heavy object onto a table.
8. Have each group share their work with the class.
9. Ask students where they have seen devices like these. Generate an experience chart.
10. Name more complex mechanisms, such as the bicycle, and ask students to tell what simple machines they might find in it.
11. Have each group examine one of the complex mechanisms and identify all of the simple machines they find in it. They should sketch some of the simple machines they find and explain what they do.

Tips

- Use familiar and unfamiliar devices so students can move from the known to the unknown and apply what they learn.
- Review rules of safety.

Worksheet for Activity #7

"Simple Machines"

Name ______________________________

Pick one simple machine. Think of a use for it showing how it could help to make a job easier. Draw or write your idea.

This is a picture of ______________________________

____

This is how this simple machine can make a job easier. ______________________________

____

____

____

____

____

EXTENSION TO

Activity №7

Looking at Larger Mechanisms

Grade Level
3-6

Overview
With this activity, students explore complex mechanisms in order to discover how subsystems contribute to larger systems.

Vocabulary
Subsystem

Skills
- Recording data
- Drawing to scale

Standards
- Standards for the English Language Arts: 12
- Benchmarks for Science Literacy: 1B, 2A, 3A, 3B
- National Science Education Standards: A, B, E

Time Needed
90 minutes

Materials
- Discarded appliances such as VCR's, tape players, rotary dial phones, typewriters
- Screwdrivers
- Large pieces of construction paper
- Pencils
- Discarded cafeteria trays or other containers for holding small parts

Procedure
1. Divide students into small groups of 3 or 4.
2. As in the Activity #7, each group should have a fairly complex mechanism.
3. Ask each group to make a diagram of their mechanism.
4. Have students in their groups remove the covers and any other parts needed in order to see the inside of their mechanism. Explain that they should remember this process because they'll need to reassemble the mechanism.
5. Ask groups to find at least one subsystem of the mechanism that they would like to investigate. For example, students may wish to examine how the stop/eject button works on a tape player or how pressing one key of a typewriter makes the type face strike the ribbon. The question should be manageable so that an investigation can take place.
6. Focusing on the question, students identify how the input leads to the output. They should describe what they find through diagrams and writing.
7. Encourage students to find and record every step in the cause-and-effect sequence leading from input to output.
8. Ask each group to record and answer questions about their discovery.
9. Bring the groups together to discuss their work.
10. Record any comments and/or questions that may arise during the discussion. These comments and/or questions may lead to other investigations.
11. Have the groups reassemble their mechanisms.

Tip
Groups may need help finding a subsystem that is not too complex or too difficult to draw and describe.

Activity № 8

How Does a Retractable Ballpoint Pen Work?

Grade Level
3-6

Prerequisites
- Knowledge of inputs/outputs
- Understanding of parts common in mechanisms

Overview
Students focus on a particular device, a retractable ballpoint pen. This activity is presented in the context of a fanciful "situational challenge."

Concepts
- Mechanisms operate based on cause-and-effect relationships among their parts.
- A mechanism cannot function properly when the cause-and-effect sequence is disrupted.

Vocabulary
- Input
- Output
- Mechanism

Skills
- Sketching
- Observing
- Sequencing cause and effect
- Writing

Standards
- Standards for the English Language Arts: 12
- Benchmarks for Science Literacy: 1B, 3A
- National Science Education Standards: A, B, E

Time Needed
45 minutes

Materials
- Worksheet #8: "The King and the Special Pens"
- See-through retractable ballpoint pens (at least 1 for every 2 students)
- Journals/notebooks
- Drawing paper

Procedure
1. Distribute Worksheet #8 to students and have them read the scenario.
2. Then demonstrate how to make the point of the pen appear and disappear without letting students see the inside of the pen.
3. Ask students to make a drawing of what they think is inside the pen based on your demonstration.
4. Ask students to think about the scenario. Ask them to look at their drawings and then to write a possible explanation for why some pens are not working.
5. Distribute pens to pairs of students.
6. Ask students to investigate what is actually inside the pen. Tell them they can disassemble the pen, but should record this process so they can reassemble the pen later.
7. Ask each pair of students to draw all of the parts of the pen and explain how each part functions to make the pen work.
8. After students have made their drawings, ask them to reassemble their pens.
9. Ask each pair to select a presenter and share their diagram.
10. Record any comments and questions on a chart tablet for further discussion. These might lead to other investigations.

Extension
Have students write an instruction book for how to assemble a ballpoint pen and use it.

Tips
- Familiarize yourself with the workings of the pen beforehand.
- Provide sufficient time and space for disassembling and reassembling the pens.

Note: *For one teacher's experience with this activity, see Chapter 4 ("Stories"), page 138.*

Worksheet for Activity #8

"The King and the Special Pens"

A king wants to free his imprisoned subjects (the fifth-grade serfs) from his dungeon. However, vandals have damaged all of the special pens used to sign the official release forms. He would like to hire someone to fix his pens so that he could sign their freedom decrees. The king will hire a subject (student) to repair the pens if s/he can convince him that s/he is a capable pen-repair technician.

Try to convince the king that you can fix his pens. You want your freedom, as well as that of all the fifth-grade prisoners. To qualify as a repairperson, you must provide:

- A written explanation of what you think could be damaged in the pens;
- A diagram of what the pen looks like inside and how it works.

You will be called to share your expertise before the king and his subjects.

Activity No 9

Make a Model of a Mechanism

Grade Level
3-6

Overview
This is a modeling activity in which students make a working model of a common mechanism from recycled newspaper and masking tape.

Prerequisite
Knowledge of slide joints, pin joints (hinges), input, output, and levers

Concept
Students will recognize mechanisms through the use of modeling

Vocabulary
- Input
- Output
- Lever
- Joint
- Pin joint
- Slide joint
- Hinge

Skills
- Observing
- Collecting data
- Drawing to scale

Standards
- Benchmarks for Science Literacy: 1B, 2C, 3B, 3C, 8B, 11B, 11D, 12C
- National Science Education Standards: A, E
- Principles and Standards for School Mathematics: C1, M1, G1, G4

Time Needed
Two 45-minute sessions (although more time may be needed, depending upon the mechanism)

Materials
- Worksheet #9: "Make a Model of a Mechanism"
- Recycled newspaper
- Masking tape
- Scissors
- Rulers
- Cardboard
- Glue
- Rubber bands
- Stapler
- Small tools (screw drivers, wire cutters, etc.)
- Oak tag, notebooks
- Pencils
- Discarded mechanisms, including at least two umbrellas (one for each group of 3-4 students)

Procedure
1. Explain to students that they will be designing a model of a mechanism using recycled materials. Use a working umbrella frame as an example.
2. Let students examine the umbrella.
3. Elicit from students their ideas about how the umbrella opens and closes. Encourage them to use vocabulary learned in prior lessons such as input, output, lever, joints, etc.
4. Record students' responses on chart paper or the board.

Note: *For a review of the ideas behind this activity,* *see Chapter 1 ("Appetizers"), page 19.*

5. Discuss the role of joints in mechanisms. Demonstrate how slide joints and pin joints (hinges) work. To demonstrate the role of joints in an umbrella, cut away the fabric from one of the umbrellas.
6. Draw the umbrella mechanism on chart paper or the board.
7. Divide students into small groups of 3 or 4.
8. Let each group choose a mechanism they would like to make a model of.
9. Distribute copies of Worksheet #9.
10. Give groups time and materials to create their models.
11. When the models are complete, have each group present its model to the whole group and describe their design process, the problems they encountered, and how they solved them.

Tips

- For further explanation of joints and how they operate, see Chapter 2 of this Guide.
- Each group should share any difficulties, questions, and comments with other groups. Time may be an issue; therefore allow enough time for the activity by limiting the types of mechanisms that students may model. Sample mechanisms may include a folding chair, office collator, vise grip, eyelash curler, tin snips, and nail clipper.

Extension

Introduce students to Rube Goldberg's inventions and let them design and create a model of one of their own.

Worksheet for Activity #9

"Make a Model of a Mechanism"

Name ______________________________

Draw and describe your model.

Activity № 10

Conductors and Insulators

Grade Level
3-6

Overview
Students test materials to determine whether or not they conduct electricity.

Prerequisites
- Understanding of insulators and conductors
- Discuss safety rules regarding electrical wires.

Concepts
- Different materials have different electrical properties
- The roles of insulators and conductors

Vocabulary
- Conductor
- Insulator
- Circuit
- Current

Skills
- Predicting outcomes
- Drawing conclusions
- Record keeping
- Collecting data

Standards
- Benchmarks for Science Literacy: 1A, 4G
- Standards for the English Language Arts: 12
- Principles and Standards for School Mathematics: DA & P1, DA & P3
- National Science Education Standards: A, B

Time Needed
45 minutes

Materials
- Worksheet #10: "Conductors and Insulators"
- Notebooks
- Pencils
- One set-up for every group of students, consisting of:
    - 3 pieces of bell wire (6" each) with ends stripped
    - Battery holder
    - Socket
    - 1.5 volt bulbs
    - Size D batteries
    - Assorted objects to test (door knob, chairs, tables, jewelry, etc.)

Procedure
1. Have one set-up displayed along with an object that insulates and one that conducts electricity.
2. Review the definitions of conductor and insulator and relate them to the examples available. (An electrical conductor allows electricity to flow through it. An insulator prevents electricity from flowing through it.)
3. Divide students into small groups.
4. Ask the groups to look around the classroom for objects to test as conductors and insulators.
5. Ask them to make predictions as to whether they believe each object is a conductor or an insulator. They should record their predictions on Worksheet #10.
6. Have each group make a circuit tester using 1 battery and holder, 1 socket and bulb, and 3 wires.
7. The groups then use their circuit testers to test their predictions, and record their findings on the Worksheet.
8. Bring the groups together to discuss their findings. Create one chart that lists all objects, the predictions, and the outcomes (insulator or conductor).

Tips
- You may want to repeat this activity, using different objects to test.
- Encourage students to use vocabulary such as conductor, insulator, circuit, current, etc.

Worksheet for Activity #10

"Conductors and Insulators"

Name ______________________________

List objects to be tested. Make predictions: Will an object conduct or insulate electricity? Then test each object.

Make a check mark under the appropriate column.

	Prediction		Actual	
Object	**Conduct**	**Insulate**	**Conduct**	**Insulate**

What other objects might conduct electricity? List them.

What other objects might insulate electricity? List them.

Activity № 11

Electric Switches

Grade Level
3-6

Overview
Students will incorporate a switch into a battery/bulb circuit. Students will also use their knowledge of circuits to design and make their own switches using common materials.

Prerequisite
Knowledge of how to create a circuit

Concept
Electric current can be controlled with a switch

Vocabulary
- Circuit
- Current
- Switch

Note: *For a review of these concepts and terms, see the Glossary at the back of this Guide.*

Skills
- Scientific testing
- Observing
- Collecting data
- Record-keeping

Standards
- Benchmarks for Science Literacy: 1B, 3A, 8C, 11A, 11B
- Standards for the English Language Arts: 12
- National Science Education Standards: A, B, E

Time Needed
45 minutes

Materials
- Worksheet #11: "Electric Switches"
- One set-up for every group consisting of the following:
  - Different types of switches
  - Wire
  - Battery holder
  - D size battery
  - Socket,
  - 1.5 volt bulb
  - Motors
  - Buzzers
  - Paper clips
  - Metal hair clips
  - Aluminum foil
  - Pliers
  - Discarded toys and appliances

Procedure
1. Ask students for examples of how switches are used. Record their responses on chart paper or the board.
2. Divide students into small groups.
3. Have students build a working circuit with just a battery and a bulb.
4. Next, ask them to add a switch in such a way that the switch turns the bulb on and off.
5. Have students draw and describe their circuits and switches on Worksheet #11.
6. Once they have been able to incorporate one type of switch, have each group create their own switch for their circuit using the materials provided.

Tips
If students have difficulty in creating a switch, suggest they use a paper clip bent so it can just barely touch a piece of foil or another paper clip when pressed; or a paper fastener wedged in between a battery and battery holder which prevents them from making contact when pressed.

Extension
Have students look for switches in discarded or appliances, toys, and other electrical devices. They can disassemble them to see how they work.

Worksheet for Activity #11

"Electric Switches"

Name ______________________________

Draw the circuits you made.

Circuit without switch	**Circuit with switch**

Describe what you did to r

What did you use to make y

Activity № 12

Two Switches, One Lamp

Grade Level
5-6

Overview
This activity starts from an everyday situation, in which one light or appliance is controlled by two switches, which both have to be on for the device to operate. Students find examples of this type of circuit, and then model it using batteries and bulbs.

Prerequisite
Knowledge of circuits, switches, conductors, and insulators

Vocabulary
- Circuit
- Conductor
- Insulator
- Switch

Note: *For a review of these concepts and terms, see "Circuit Situations Revealed" in Chapter 2.*

Skills
- Mapping
- Recording
- Observing
- Investigating

Standards
- Benchmarks for Science Literacy: 1A, 1B, 3A, 8C, 11A, 11B
- Standards for the English Language Arts: 12
- Principles and Standards of School Mathematics: A1, A3, C3
- National Science Education Standards: A, B, E

Time Needed
Two 45-minute sessions.

Materials:
- Worksheet #12: "Two Switches, One Lamp"
- One set-up for each group consisting of the following:
    - D size battery
    - 1.5 volt bulb and socket
    - Bell wire
    - Battery holder
    - Two switches (ready made or home made)
- Small lamp
- Computer power strip

Procedure
1. Show students the small lamp. Plug it into the computer power strip and plug the power strip into the wall. Ask if they can name two ways to turn the lamp on and off. If necessary, demonstrate for class. Anything plugged into it will turn on only if its own switch and the power strip switch are both on.
2. Divide students into groups providing one set-up per group.
3. Present them with the following challenge: Can you design a circuit that contains two switches, both of which control the flow of electricity to one bulb? Just as with the power strip and lamp switch, both switches have to be on for the bulb to come on.
4. Each group should draw a diagram of the circuit they think would work, and present it to the class.
5. Once each group has a viable diagram, students begin to construct their circuits with switches.
6. If any of the groups are having trouble, ask, "If either switch could stop the bulb from lighting, where would the switches have to be? If only one switch was open, would the other one allow the current to light the bulb?"
7. Each group shares their design after construction.

Tips
- Allow plenty of time for each group to complete the assignment.
- Encourage students to visit other groups and brainstorm ideas.

Worksheet for Activity #12

"Two Switches, One Lamp"

Name

Draw your circuit design.

Did your circuit/switches model your design?

Did you modify your design? If so, how?

How did your design work?

Did your design succeed or fail? Why?

Activity № 13

Electric Circuit Board Game

Grade Level
5-6

Overview
Students use their knowledge of circuits to design and make an electric question-and-answer board game.

Prerequisite
Knowledge of circuitry

Concepts
Electric current behavior

Skills
- Drawing conclusions
- Modeling
- Designing

Standards
- Benchmarks for Science Literacy: 8C, 11A, 11B
- Standards for the English Language Arts: 12
- National Science Education Standards: A, B, E

Time Needed
Two or three 45-minute sessions

Materials
- Worksheet #13: "Electric Circuit Board Game"
- Oak tag
- Cardboard
- Markers
- Construction paper
- Aluminum foil
- Hole puncher
- Masking tape
- Paper fasteners
- Circuit testers (batteries, bulbs, wires)
- Large heavy duty rubber bands (to hold the game together)

Procedure
1. Explain to students that they will be creating a game that will indicate right or wrong answers to questions.
2. Divide students into small groups.
3. Have students come up with questions and answers within their favorite curriculum areas to be used in their games.
4. Each group then designs a model of a game board on paper.
5. Here's how to construct the game:
   - On their cardboard game boards, students use pieces of aluminum foil to connect a small hole next to each question with another hole next to its correct answer. Masking tape should be used to insulate the aluminum foil, especially where two pieces of aluminum foil have to cross.
   - Students use oak tag with pre-cut holes to cover the circuitry, so that players can't see the aluminum foil connecting the questions and answers. The questions can be written directly on the oak tag.
6. The game is played using the circuit tester. (See Activity #10.) The player places 1 wire of the circuit tester in a hole next to the question, then places the 2nd wire in the hole next to a possible answer. If the answer is correct, the bulb will light.
7. Let groups take turns playing each others' games.

Tips
- Review safety rules.
- It is easier to construct the circuit board first.
- Use heavy-duty aluminum foil.
- Model how to fold the aluminum and cover it with masking tape. Alternatively, wire and paper fasteners can be substituted for the aluminum foil.

Worksheet for Activity #13

"Electric Circuit Board Game"

Name ______________________________

Draw your game board design.

What problems came up when designing your game board?

____

____

What can you change in your design to make it better?

____

____

Activity No 14

Water-Level Alarm

Grade Level
5-6

Overview
Students design and construct a float-controlled electric alarm system that lights a bulb or activates a buzzer when the water rises above a certain level. The float carries an automatic switch contact that automatically completes a circuit when it reaches the appropriate level.

Prerequisite
Knowledge of designing a circuit

Concept
Electric current behavior

Skills
- Measuring
- Estimating
- Designing

Standards
- Benchmarks for Science Literacy: 8C, 11A, 11B
- Standards for the English Language Arts: 12
- National Science Education Standards: A, B, E

Time Needed
Two or three 45-minute sessions

Materials
- Worksheet #14: "Water-Level Alarm"
- Plastic basin
- Water
- Insulated wire
- D size batteries
- Battery holder
- 1.5 volt bulbs or buzzers
- Sockets
- Pitchers
- Aluminum foil
- Pencils
- Paper
- Measuring cups
- Scissors
- Floats (styrofoam or wood)
- Screwdrivers
- Nails
- Funnels
- Rulers

Procedure
1. Review safety and clean-up rules concerning water.
2. Describe this scenario to students: "You have been asked to design a water-level alarm for the sight- or hearing-impaired. It will be used in the bathroom to let a person know when their tub is full."
3. Divide students into small groups and distribute copies of Worksheet #14.
4. Ask each group to plan and draw their designs before starting construction.
5. Review circuits by having the students construct a simple circuit with a switch.
6. You may want to model several types of floats using styrofoam or wood.
7. If students have difficulties, facilitate their work with questions and design tips.
8. Have the groups share their designs with the class and describe their process, problems, and solutions.

Tip
This lesson can be used as an assessment of Activities 10-13.

Worksheet for Activity #14

"Water-Level Alarm"

Name ______________________________

Draw and describe your water-level alarm.

Is this water level alarm made for a sight- or hearing-impaired person?

How would you change the design to make it better?

Standards for Activities

Activity #1: What Is a Mechanism?

Standards for the English Language Arts
Standard #12: Students use spoken, written and visual language to accomplish their own purposes.

National Science Education Standards
Content Standard A: Students should develop abilities to do scientific inquiry.

Benchmarks for Science Literacy
Benchmark #1B: Describing things as accurately as possible is important in science because it enables people to compare their observations with those of others.

Activity #2: Be a Mechanism Detective

Standards for the English Language Arts
Standard #12: Students use spoken, written, and visual language to accomplish their own purposes.

National Science Education Standards
Content Standard A: Students should develop abilities to do scientific inquiry.

Principles and Standards for School Mathematics
Geometry Standard G4: Use visualization, spatial reasoning, and geometric modeling to solve problems.

Benchmarks for Science Literacy
Benchmark #1B: Scientific investigations may take many different forms, including observing what things are like or what is happening somewhere, collecting specimens for analysis, and doing experiments.
Benchmark #2A: Mathematics is the study of many kinds of patterns. Patterns are studied because they help to explain how the world works or how to solve practical problems.
Benchmark #2C: Mathematicians often represent things with abstract ideas, such as numbers or straight lines.

Activity #3: What Does a Tool Do?

Standards for the English Language Arts
Standard #12: Students use spoken written language to accomplish their own purposes.

National Science Education Standards
Content Standard A: Students should develop abilities to do scientific inquiry.
Content Standard B: Students should develop an understanding of position and motion of objects.

Principles and Standards for School Mathematics
Geometry Standard G1: Analyze characteristics and properties of two- and three-dimensional geometric shapes and develop mathematical arguments about geometric relationships.

Benchmarks for Science Literacy
Benchmark #1B: People can often learn about things around them by just observing those things carefully, but sometimes they can learn more by doing something to the things and noting what happens.
Benchmark #2A: Things move, or can be made to move, along straight, curved, circular, back-and-forth, and jagged paths.
Benchmark #3A: Tools are used to do things better or more easily and to do some things that could not otherwise be done at all.

Activity #4: Can You Guess My Categories?

Standards for the English Language Arts
Standard #7: Students conduct research on issues. They gather, evaluate, and synthesize data from a variety of sources to communicate their discoveries in ways that suit their purpose and audience.
Standard #12: Students use spoken, written, and visual language to accomplish their own purposes.

National Science Education Standards
Content Standard A: Students should develop abilities to do scientific inquiry.

Principles and Standards for School Mathematics
Algebra Standard A1: Understand, patterns, relations, and functions.

Benchmarks for Science Literacy
Benchmark #1B: Scientific investigations may take many different forms, including observing what things are like or what is happening somewhere, collecting specimens for analysis, and doing experiments.
Benchmark #1C: In doing science, it is often helpful to work with a team and to share findings with others.
Benchmark #2C: Numbers and shapes—and operations on them—help to describe and predict things about the world around us.

Activity #5: Ins and Outs of Inputs and Outputs

Standards for the English Language Arts
Standard #12: Students use spoken, written, and visual language to accomplish their own purposes.

National Science Education Standards
Content Standard A: Students should develop abilities to do scientific inquiry.
Content Standard B: Students should develop an understanding of position and motion of objects.

Principles and Standards for School Mathematics
Geometry Standard G1: Analyze characteristics and properties of two- and three-dimensional geometric shapes and develop mathematical arguments about geometric relationships.

Benchmarks for Science Literacy
Benchmark #1B: Results of scientific investigations are seldom exactly the same, but if the differences are large, it is important to try to figure out why.
Benchmark #3A: Technology enables scientists and others to observe things that are too small or too far away to be seen without them and to study the motion of objects that are moving very rapidly or are hardly moving at all.
Benchmark #3B: There is no perfect design. Designs that are best in one respect may be inferior in other ways.
Benchmark #11A: In something that consists of many parts, the parts usually influence one another.

Activity #6: How Do Levers Make Work Easier?

Standards for the English Language Arts
Standard #12: Students use spoken, written, and visual language to accomplish their own purposes.

National Science Education Standards
Content Standard A: Students should develop abilities to do scientific inquiry.
Content Standard B: Students should develop an understanding of position and motion of objects.

Principles and Standards for School Mathematics
Algebra Standard A1: Understand, patterns, relations, and functions.
Data Analysis and Probability Standard DA & P 1: Formulate questions that can be addressed with data and collect, organize, and display relevant data to answer them.
Data Analysis and Probability Standard DA & P 3: Develop and evaluate inferences and predictions that are based on data.
Connections Standard C3: Recognize and apply mathematics in contexts outside of mathematics.
Measurement Standard M1: Understand measurable attributes of objects and the units, systems, and processes of measurement.

Benchmarks for Science Literacy
Benchmark #1B: Results of scientific investigations are seldom the same, but if different it is important to figure out why. Following directions and keeping records of work is a way to provide information on what might have caused the differences.
Benchmark #2A: Mathematics is the study of many kinds of patterns, including numbers and shapes and operations on them.
Benchmark #3A: Measuring instruments can be used to gather accurate information for making scientific comparisons of objects and events and for designing and constructing things that will work properly.

Activity #7: Simple Machines

Standards for the English Language Arts
Standard #12: Students use spoken, written, and visual language to accomplish their own purposes.

National Science Education Standards
Content Standard A: Students should develop abilities to do scientific inquiry.
Content Standard B: Students should develop an understanding of position and motion of objects.
Content Standard E: Students should develop understanding about science and technology.

Benchmarks for Science Literacy
Benchmark #1B: Scientists' explanations about what happens in the world come partly from what they observe, partly from what they think.
Benchmark #2A: Mathematics is the study of many kinds of patterns, including numbers and shapes and operations on them. Sometimes patterns are studied because they help to explain how the world works or how to solve practical problems.

Activity #7 Extension: Looking at Larger Mechanisms

Standards for the English Language Arts
Standard #12: Students use spoken, written, and visual language to accomplish their own purposes.

National Science Education Standards
Content Standard A: Students should develop abilities to do scientific inquiry.
Content Standard B: Students should develop an understanding of position and motion of objects.
Content Standard E: Students should develop understanding about science and technology.

Benchmarks for Science Literacy
Benchmark #1B: Scientists' explanations about what happens in the world come partly from what they observe, partly from what they think.
Benchmark #2A: Mathematics is the study of many kinds of patterns, including numbers and shapes and operations on them. Sometimes patterns are studied because they help to explain how the world works or how to solve practical problems.
Benchmark #3A: Technology enables scientists to study the motion of objects that are moving very rapidly or are hardly moving at all.
Benchmark #3B: The solution to one problem may create other problems.

Activity #8: How Does a Retractable Ballpoint Pen Work?

Standards for the English Language Arts

Standard #12: Students use spoken, written, and visual language to accomplish their own purposes.

Benchmarks for Science Literacy

Benchmark #1B: Scientific investigations may take many different forms, including observing what things are like or what is happening somewhere.

Benchmark #3A: Throughout all of history, people everywhere have invented and used tools. Most tools of today are different from those of the past but are modifications of ancient tools.

National Science Education Standards

Content Standard A: Students should develop abilities to do scientific inquiry.

Content Standard B: Students should develop an understanding of position and motion of objects.

Content Standard E: Students should develop understanding about science and technology.

Activity #9: Make a Model of a Mechanism

Benchmarks for Science Literacy

Benchmark #1B: Scientists do not pay much attention to claims about how something they know about works unless the claims are backed up with evidence that can be confirmed and with a logical argument.

Benchmark #2C: In using math, choices have to be made about what operations will give the best results. Results should always be judged by whether they make sense and are useful.

Benchmark #3B: There is no perfect design.

Benchmark #3C: Any invention is likely to lead to other inventions. Once an invention exists, people are likely to think up ways of using it that were never imagined at first.

Benchmark #8B: Discarded products contribute to the problem of waste disposal. Sometimes it is possible to use the materials in them to make new products, but materials differ widely in the ease with which they can be recycled.

Benchmark #11B: Seeing how a model works after changes are made to it may suggest how the real thing would work if the same were done to it.

Benchmark #11D: Almost anything has limits on how big or small it can be.

Benchmark #12C: Choose appropriate common materials for making simple mechanical constructions and repairing things.

National Science Education Standards

Content Standard A: Students should develop abilities to do scientific inquiry.

Content Standard B: Students should develop an understanding of position and motion of objects.

Content Standard E: Students should develop understanding about science and technology.

Principles and Standards for School Mathematics

Connections Standard C3: Recognize and apply mathematics in contexts outside of mathematics.

Measurement Standard M1: Understand measurable attributes of objects and the units, systems, and processes of measurement.

Geometry Standard G1: Analyze characteristics and properties of two- and three-dimensional geometric shapes and develop mathematical arguments about geometric relationships.

Geometry Standard G4: Use visualization, spatial reasoning, and geometric modeling to solve problems.

Activity #10: Conductors and Insulators

Benchmarks for Science Literacy
Benchmark #1A: Results of similar scientific investigations seldom turn out exactly the same. Sometimes this is because of unexpected differences in the things being investigated.
Benchmark #4G: Different kinds of materials respond differently to electric forces. In conducting materials such as metal, electric charges flow easily, whereas in insulating materials such as glass, they can move hardly at all.

Standards for the English Language Arts
Standard #12: Students use spoken, written, and visual language to accomplish their own purposes.

National Science Education Standards
Content Standard A: Students should develop abilities to do scientific inquiry.
Content Standard B: Students should develop an understanding of light, heat, electricity and magnetism.

Principles and Standards for School Mathematics
Data Analysis and Probability Standard DA & P 1: Formulate questions that can be addressed with data and collect, organize, and display relevant data to answer them.
Data Analysis and Probability Standard DA & P 3: Develop and evaluate inferences and predictions that are based on data.

Activity #11: Electric Switches

Benchmarks for Science Literacy

Benchmark #1B: Scientific investigations may take many different forms including observing what things are like or what's happening somewhere, collecting specimens for analysis, and doing experiments.

Benchmark #3A: When trying to build something or to get something to work better, it usually helps to follow directions if there are any or to ask someone who has done it before for suggestions.

Benchmark #8C: Electrical energy can be produced from a variety of energy sources and can be transformed into almost any other form of energy. Moreover electricity is used to distribute energy quickly and conveniently to distant locations.

Benchmark #11A: Something may not work as well (or at all) if a part of it is missing, broken, worn out, mismatched, or misconnected.

Benchmark #11B: Seeing how a model works after changes are made to it may suggest how the real thing would work if the same were done to it.

Standards for the English Language Arts

Standard #12: Students use spoken, written, and visual language to accomplish their own purposes.

National Science Education Standards

Content Standard A: Students should develop abilities to do scientific inquiry.

Content Standard B: Students should develop an understanding of light, heat, electricity and magnetism.

Content Standard E: Students should develop understanding about science and technology.

Activity #12: Two Switches, One Lamp

Benchmarks for Science Literacy

Benchmark #1A: Results of similar scientific investigations seldom turn out exactly the same. Sometimes this is because of unexpected differences in the things being investigated.

Benchmark #1B: What people expect to observe often affects what they actually do observe. Strong beliefs about what should happen can prevent them from detecting other results. One safeguard is to have different investigations conduct independent studies of the same questions.

Benchmark #3A: Engineers, architects, and others who engage in design and technology use scientific knowledge to solve practical problems, but they usually have to take human values and limitations into account as well.

Benchmark #8C: Different ways of obtaining, transforming, and distributing energy have different environmental consequences.

Benchmark #11A: Thinking about things as systems means looking for how every part relates to others. The output from one part of a system (which can include material, energy or information) can become the input to other parts. Such feedback can serve to control what goes on in the system as a whole.

Benchmark #11B: Models are often used to think about processes that happen too slowly, too quickly, or on too small a scale to observe directly, or that are too vast to be changed deliberately, or that are potentially dangerous.

Standards for the English Language Arts

Standard #12: Students use spoken, written, and visual language to accomplish their own purposes.

National Science Education Standards

Content Standard A: Students should develop abilities to do scientific inquiry.

Content Standard B: Students should develop an understanding of light, heat, electricity and magnetism.

Content Standard E: Students should develop abilities of technological design and understanding about science and technology.

Principles and Standards for School Mathematics

Algebra Standard A1: Understand, patterns, relations, and functions.

Algebra Standard A3: Use mathematical models to represent and understand quantitative relationships.

Connections Standard C3: Recognize and apply mathematics in contexts outside of mathematics.

Activity #13: Electric Circuit Board Game

Benchmarks for Science Literacy
Benchmark #8C: Electrical energy can be produced from a variety of energy sources and can be transformed into almost any other from of energy. Moreover, electricity is used to distribute energy quickly and conveniently to distant locations.
Benchmark #11A: A system can include processes as well as things.
Benchmark #11B: Different models can be used to represent the same things.

Standards for the English Language Arts
Standard #12: Students use spoken, written, and visual language to accomplish their own purposes.

National Science Education Standards
Content Standard A: Students should develop abilities to do scientific inquiry.
Content Standard B: Students should develop an understanding of light, heat, electricity and magnetism.
Content Standard E: Students should develop abilities of technological design and understanding about science and technology.

Activity #14: Water-Level Alarm

Benchmarks for Science Literacy
Benchmark #8C: Electrical energy can be produced from a variety of energy sources and can be transformed into almost any other form of energy. Moreover, electricity is used to distribute energy quickly and conveniently to distant locations.
Benchmark #11A: A system can include processes as well as things.
Benchmark #11B: Different models can be used to represent the same thing.

Standards for the English Language Arts
Standard #12: Students use spoken, written, and visual language to accomplish their own purposes.

National Science Education Standards
Content Standard A: Students should develop abilities to do scientific inquiry.
Content Standard B: Students should develop an understanding of light, heat, electricity and magnetism.
Content Standard E: Students should develop abilities of technological design and understanding about science and technology.

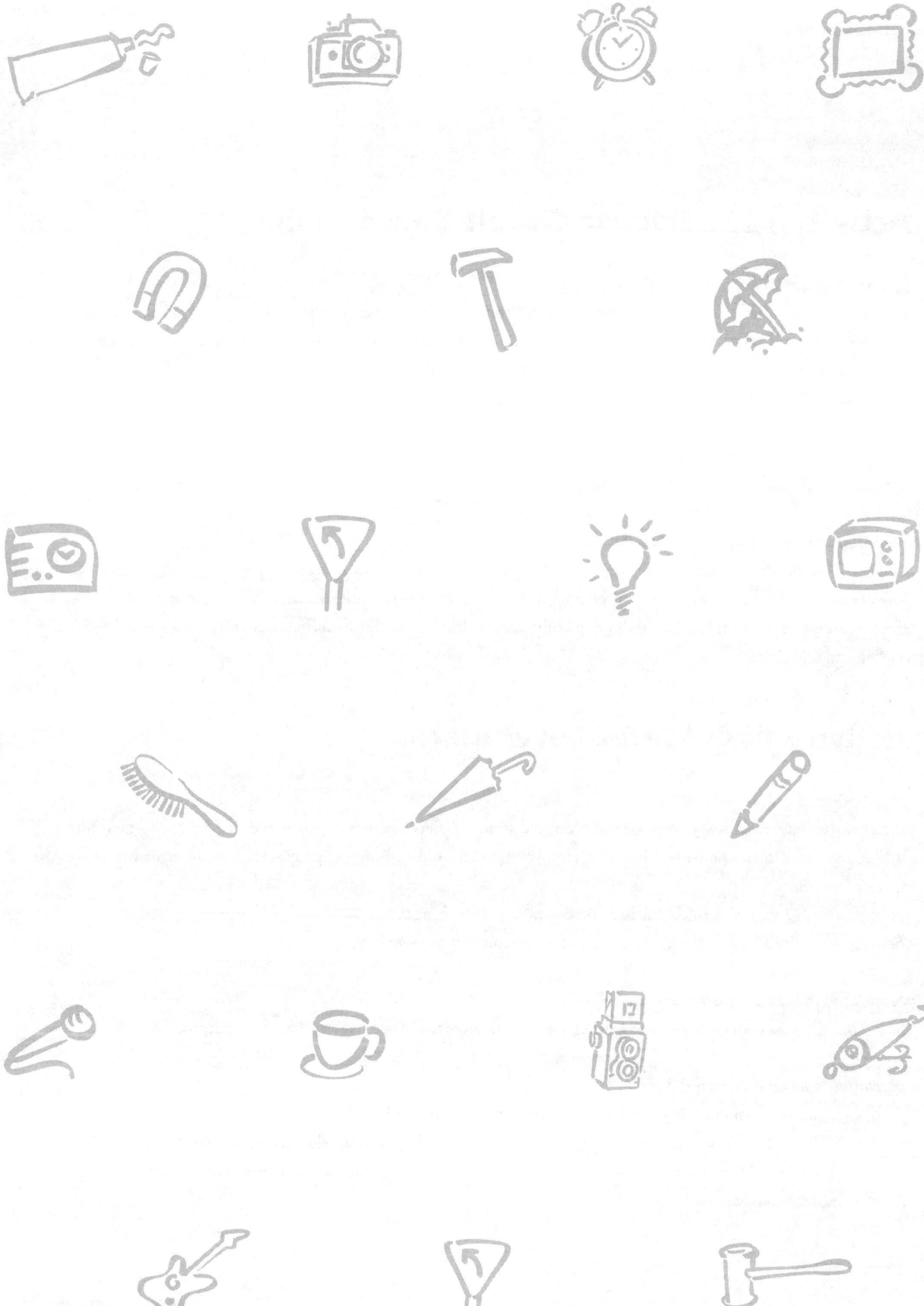

Chapter 4

Stories

Mechanisms have immediate appeal to nearly everyone who examines them, and they can be approached at many different levels. Table 4-1 suggests some of the ways to approach the study of mechanisms, from the most elementary to the most challenging.

In this chapter, five elementary teachers tell the stories of how they approached the study of mechanisms in their classrooms. These teachers worked at a variety of grade levels, and toward a variety of goals. Some saw mechanisms as a vehicle for teaching language arts, while others were more concerned with science concepts, such as levers and simple machines. Each began with one or more of the "Getting Started" activities and proceeded to activities from the "Analysis" and/or "Design" categories. As you read through their accounts, you will see a variety of ways of structuring this progression.

The study of circuits can also be approached first through hands-on experiments, scavenger hunts, and classifying. However, as noted in Chapter 2, circuits cannot be analyzed in quite the same way as mechanisms because the moving parts (electrons) can be neither seen nor felt. On the other hand, modeling and design activities with batteries, bulbs, and switches quickly reveal what it takes to make a circuit work. The section on circuits features two teachers whose stories about teaching mechanisms are also included in this chapter. The integrative theme for both of these teachers was "controls." Their students came to see how both mechanical and electrical controls share the common property of "influencing something outside of its own structure."

Table 4-1

Approaches to the study of mechanisms

Getting Started

Hands-on manipulation and sketching

Scavenger hunts

Sorting and classifying

Analysis

Looking for simple machines

Identifying effort, fulcrum and load

Finding first, second and third class levers

Design

Modeling existing mechanisms using recyled materials

Concept design of a mechanism to accomplish a particular task

Design and testing of pop-ups and toys

Five Teachers' Stories About Teaching Mechanisms

Annette Purnell, a first-grade teacher, used the study of mechanisms as a vehicle for developing her students' intellectual and social awareness. Her unit begins with analysis of the motions of simple tools and ends with designing and testing working models of scissors.

Kathy Aguiar, a bilingual special education teacher, began with a vocabulary question: "What is a control?" After doing some scavenger hunts, her children identified the control devices on familiar appliances and wrote about what the controls do. Their work with controls helped them overcome some of their inhibitions towards writing.

Shirley Monterroso-Nieves, a second-grade teacher with little previous experience in science or technology, engaged her students in an extended unit on mechanisms. They began by trying to explain what the word "mechanism" might mean. The students looked at simple devices to determine what they do and how they move. By the end of the unit, the children were using the terms "fulcrum," "effort," and "load" in analyzing these devices.

Mary Flores, a resource room (special education pullout) teacher engaged her fourth and fifth graders in both analysis and design of mechanisms. They began by trying to define "mechanism," by exploring simple devices, and by categorizing them. Eventually, they were able to design, evaluate, and redesign complex contraptions such as windlasses and "book-turning machines."

Angel Gonzalez is a science specialist. He introduced the work with mechanisms by setting up an imaginary situation involving a king, prisoners, and evil vandals. Within a short time, his students were able to describe the operation of some fairly complex devices: a retractable ballpoint pen and a valve for controlling water supply. This unit on mechanical controls led directly to further work with electrical switches.

The First-Grade "Mechanism Detectives" *by Annette Purnell*

Annette Purnell has taught kindergarten and first grade at C.E.S. 42 in the South Bronx, New York, for many years. She is deeply concerned with both the social and the cognitive development of each child, which she sees as closely connected. Annette began her work with mechanisms by using the word "tool" as the stimulus for an experience chart.

I asked a question: "What is a tool?" I wanted to find out how familiar that term was to my first-graders. Several children could name tools and many could not. Then as the children began to describe tools they were familiar with, others raised their hands or called out to contribute. Afterwards, we sat on the perimeter of the rug and passed around a one-hole punch to explore how it worked. The class watched as each child demonstrated it. One child traced the hole punch in the rest position. Another child traced it in its contracted position. When the hole punch was in its contracted position, two people had to help each other.

After this initial experience, Annette wanted her students to examine a variety of mechanisms closely, both to see how they work and to figure out what they are used for. She called this activity "Be a Mechanism Detective."

The children met as a whole class. I asked them to divide into groups of three or four. I kept six children, those who need much scaffolding and modeling before they can be independent. I became part of their group.

Annette provided a variety of mechanisms for each group to choose from. She deliberately selected mechanisms that would probably not be very familiar to the children, so it would be a puzzle to figure them out. They included a nutcracker, a garlic press, a pair of ice tongs and a pantograph. The latter device is used to reduce or enlarge drawings.

The pantograph group was the group I worked with. I could not hear the other groups' talk since I was with a group. I noticed that two of the six children in my group became much more interested than the other four. They were excited about explaining how the pantograph worked. They may become a group of two in later work. They don't ordinarily choose each other, so it was good that they worked together.

Annette asked each student to write down how he or she thought their group's mechanism worked, and what it might be used for. One student's work is shown in Figure 4-1.

Close examination of mechanisms helps children to develop their awareness of cause-and-effect relations. Sharing is essential, either through reporting or writing. Shared work really helps to develop an awareness of audience. Written work forces the writer to organize and develop his or her thoughts.

Listening to their language is one of my primary ways of processing what they have learned. When I sit with one group I can't float. This is worrisome for me, because then I cannot hear the other groups' language until sharing time. Sometimes their reasoning and insights can be brilliant.

Forty-five minutes seems to be a good amount of time for this work, including both exploring and writing. Children who write more slowly don't always get a chance to finish. After thirty minutes of exploration, I call for them to look for a "stopping place" so they will have time to complete at least one sentence.

I see children in the same group writing the same observations. This is fine with me, because they are using one another for support. As they become more comfortable, their independence will flourish. I'll have them read their work to each other as a stimulus. I can also have each one write about a different aspect of the mechanism, which will vary the information of each child and encourage the socialization for listening.

Annette followed these "Getting Started" activities with in-depth investigations of a pair of scissors. She wanted her children to look very carefully at this already familiar device, to notice its patterns and motions and to make a paper model of it.

Before they started tracing, I led a brief discussion on looking for shapes. The children saw a triangle inside the two blades when they were in the open position. One child saw a triangle between the blade and handle on the left side and the same thing on the right side when the scissors were open. Another child described the holes within the handles as "circles."

Their next task was to trace the paths of the blades with their pencils. I modeled the method, using my

4-1: **One mechanism detective's idea about a pair of ice tongs**

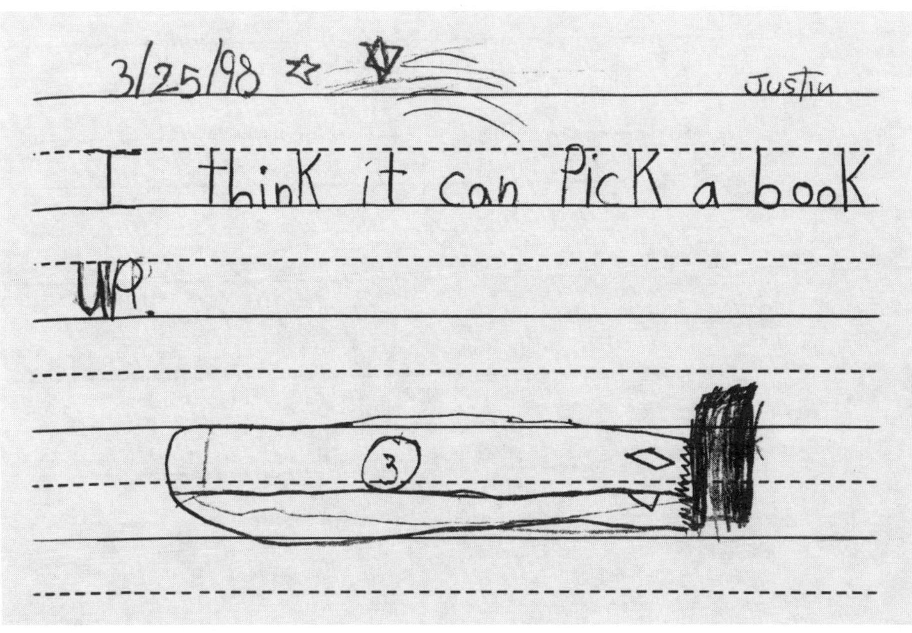

finger to simulate the pencil. Each group of two had a single pair of scissors. Having only one scissors between them presented the opportunity for each student to watch, while the other was doing. There was another advantage in working in groups of two. When a difficulty arose, the need to solve it together reinforced their reading group theme on "What friends do and how they behave."

The tracings varied. Some children traced the entire pair of scissors first, while others traced just the path. Bryant's tracing of his scissors is shown in Figure 4-2, and his description of what it does is in Figure 4-3. Latisha explains how to use a scissors in Figure 4-4.

Difficulties arose when children attempted to follow the path and hold the scissors at the same time. It turned out to be impossible to do the whole job alone. When I saw this happening, I said, "I'm seeing your scissors moving. How can your partner help?" The "tracers" knew their difficulties, so they were able to get help. The results were paths that looked more like arcs. Those who went at it alone made paths that were straight lines. At the end of the activity, we came together to discuss it. I asked who found this task hard and who found it easy. For those who found it difficult, I asked what would have helped them. Several children found the movement of the scissors confusing. I would change the presentation by suggesting that they keep one part of the scissors fixed. The stationary part would make a better frame of reference for seeing the movement of the other part.

Annette's concluding activity involved modeling. Her children had to create a cardboard model of a pair of scissors. Annette pre-cut strips of oak tag, and also provided each group with paper fasteners and a one-hole punch. They had to figure out the rest. Some of the problems were:

- Deciding how to align the strips;
- Realizing that they needed to make holes;
- Figuring out how to use the one-hole punch to make the holes;
- Deciding where to put the holes;
- Determining how to use the paper fastener and how to secure it.

4-2: **Bryant's tracing of a scissors**

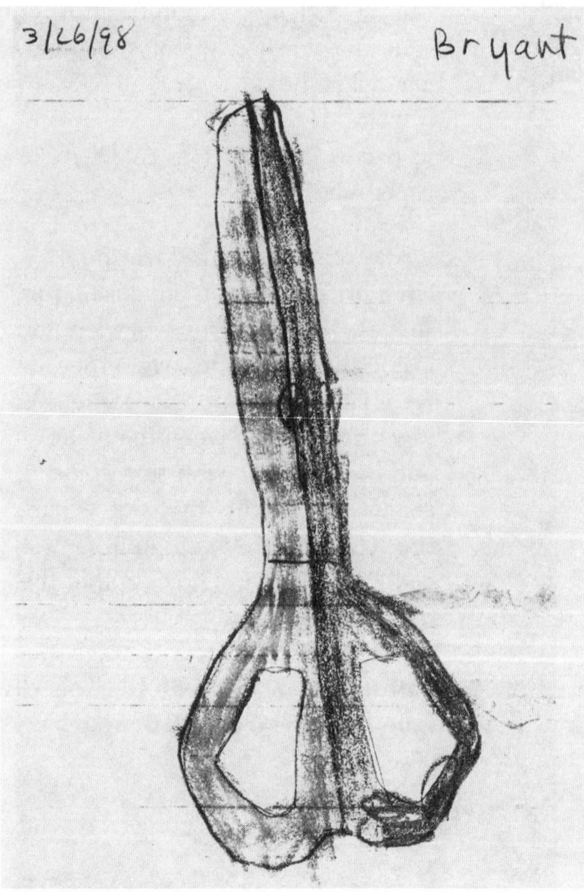

4-3: **Bryant's explanation of a scissors**

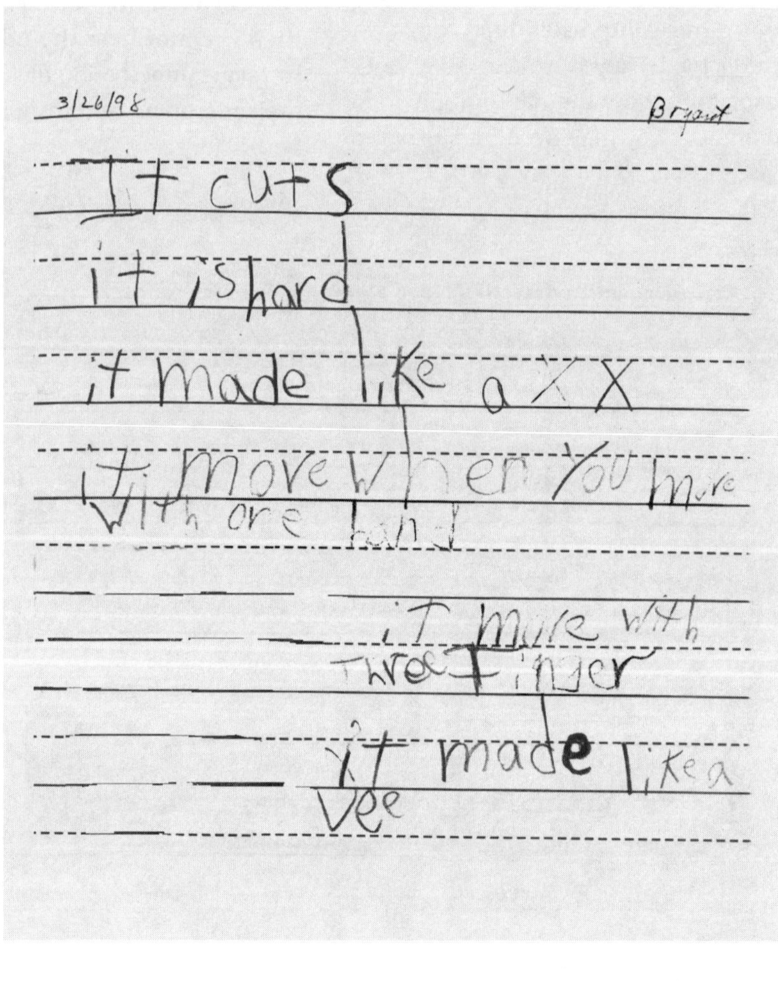

While these steps might seem obvious to an adult, they require considerable thought and discussion among first-graders. The use of the hole punch was particularly interesting to them. Although they had examined them as part of their previous explorations, they had never actually used one to make a hole.

Figure 4-5 shows the scissors model Carlos made.

Here are some of the children's comments about the activity:

MARK:
The hole punch makes a hole. I put the paper fastener through the hole.

ARLEEN:
Leroy put the strips together. Then he put the paper fastener through the hole. Now we are going to see what else it can do.

STACEY/DOMINIQUE:
We put the paper like scissors. We used our fingers to see how the strips could do the same like our hands.

RASHAUN:
I put one strip on top of the other like an "X." I punched a hole in it with a hole punch. I put the paper fastener in it. Then I put the wings of the paper fastener down on the paper.

ANTONIA:
I put the strips together so it could look like an "X." Then Crystal asked for the hole punch and she made a hole. Then we tried to cut paper with the model scissors but it didn't work. We made it open and close. We could make it turn around and around like a toy.

4-4: **How to use a scissors**

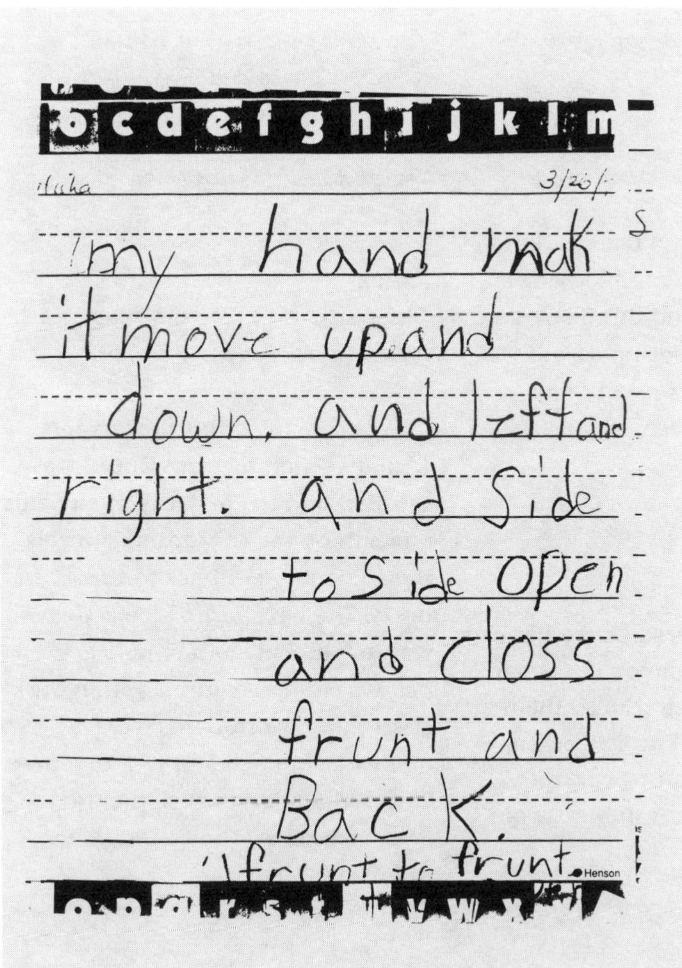

4-5: **Carlos's model of a scissors**

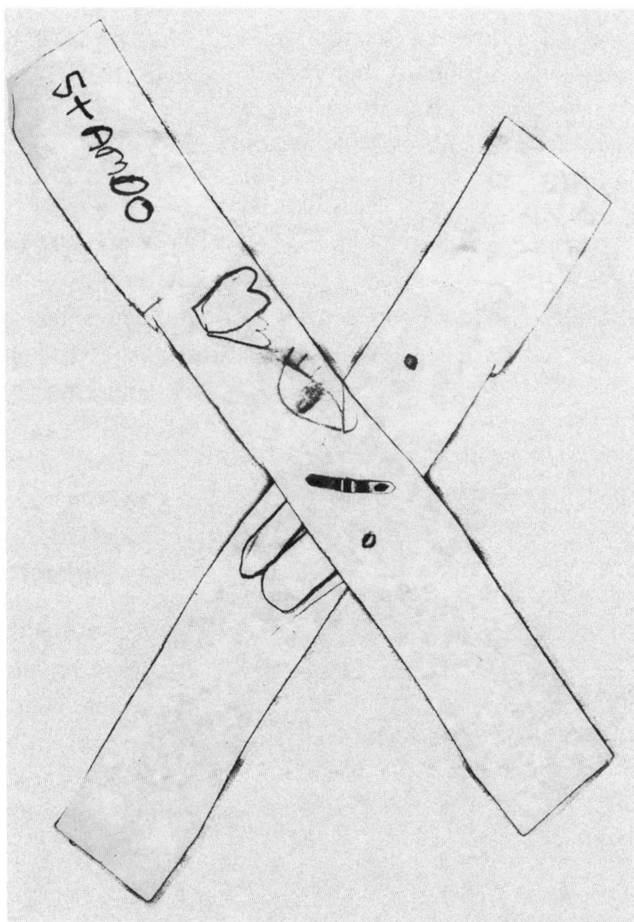

Overcoming Barriers by Writing About Controls *by Kathy Aguiar*

Kathy Aguiar taught a third-grade bilingual special education class in the same school as Annette. She began the 1998-99 school year with some serious problems.

I was greeted with 15 bright-eyed bilingual special needs students. These students had a wide range of academic as well as emotional needs. Some of my students could be very aggressive. They would throw blocks and/or chairs. All of them showed signs of frustration while writing and reading. They also had very low self-esteem. The words most often spoken by them were:

- **"I can't."**
- **"I don't know."**
- **"I can't read this."**
- **"This is too much to write."**

Kathy's challenge was to convince these students that they could actually learn to read and write and that their participation in the process was essential. She selected controls as one of her major topics. The first activity was a scavenger hunt: *Find control devices in the classroom.*

Before my students were to find controls they first had to know what a control was. I thought that I could have several controls in the middle of the rug. Students were sitting around the objects on the rug. I posed the question: "What is a control?"

JOAN:
Nintendo—the buttons on the control makes the car move.

FREDDY:
La bateria controla el reloj (The battery controls the watch).

JOSE:
On the cable box, the buttons control the channels.

YOKASTELYN:
The steering wheel moves the car to the right or left.

CARLOS:
The key lets the door open and close.

MARINA:
The pole controls the window.

We finally came to the conclusion that a control lets something else happen. While having this discussion, the students kept looking at and touching the objects on the rug, which were as follows:

- **a pen**
- **a wrench**
- **a radio**
- **a calculator**

The students were asked what the items had in common. They had not a clue. Then I introduced them to the concept of controls. I asked, "How or what is a control on the radio?" Some of the responses were as follows:

JENNIFER:
The button for the volume. It goes up and down.

FREDDY:
The button that turns it on/off.

JOAN:
The plug. If it is not plugged in then the radio won't work.

We continued this process for each object on the rug. Once I was satisfied that the majority of students understood the concept of controls, the students went back to their desks. The next part of the activity was for students to pair up. Each pair was to find controls within the classroom. The students were motivated and worked well together. They did not argue and tried to help each other. The results were as follows:

JOSE:
La pega junta dos papeles (The glue joins two papers).

ANDREW:
The batteries control the clock

JENNIFER:
The box holds the crayons.

Each student was able to explain how an item he or she drew was a control. They did become frustrated when it was time to write about the control. I did not obligate the students to write. Some children chose to write, while others simply drew and colored. (See Figure 4-6.)

In her reflections on this activity, Kathy recognized that the children used the word "control" in its everyday sense rather than its technical sense. They considered anything a "control" that causes another thing to happen. In its technical meaning, discussed in Chapter 2, the radio buttons would be considered controls, but not the plug, glue, box, or batteries. Only the buttons use minimal energy to control larger flows of energy elsewhere.

Nevertheless, Kathy was satisfied with this activity, because her primary goals were to engage the children socially and intellectually, and in ways they would not find threatening. Because they are familiar, interesting and fun, controls had proven to be an excellent vehicle for accomplishing these goals.

4-6: **Looking for the control on a can opener**

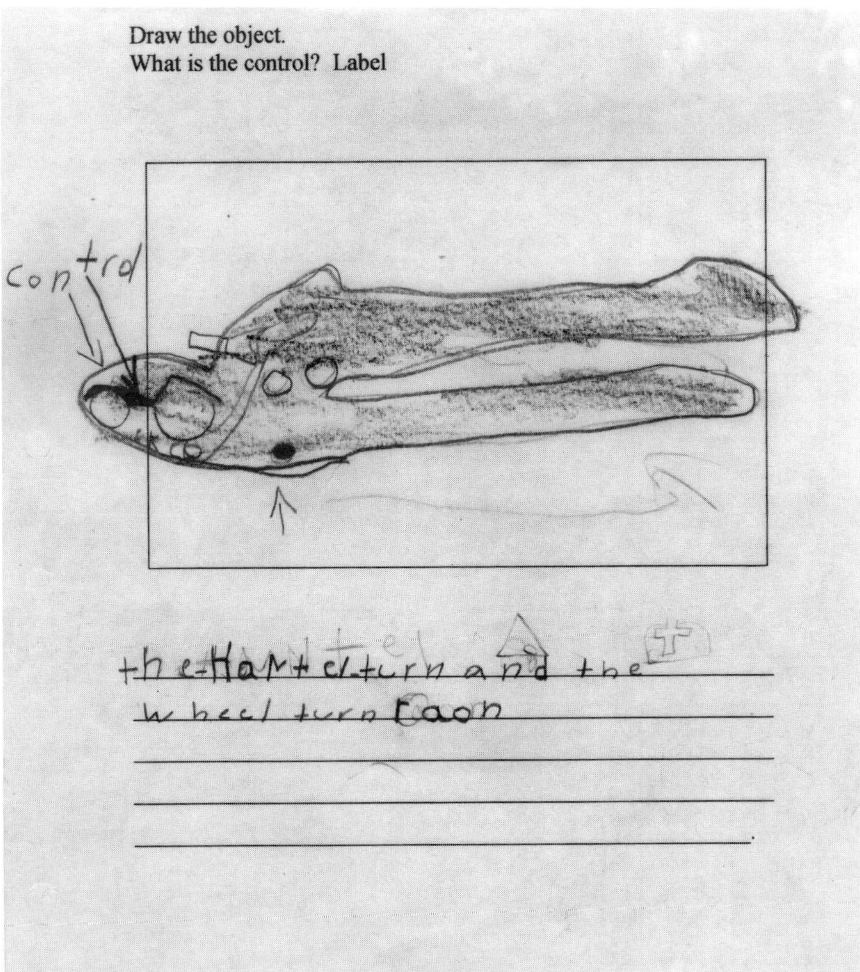

The next activity was to have students go around the school looking for controls. They were to record their findings on an activity sheet. When the activity was explained to the students, they were very willing to complete it. While planning this activity, however, I was concerned about several students' behavior. Once we were ready to go into the hallway, one of my students decided not to go. It took more than 15 minutes to finally get him into the hallway with the rest of the class. I began to think that this activity might not have been a good idea just yet. However, the rest of the class was very well motivated. They followed rules and looked for controls in an orderly manner.

We began this activity by looking at the classroom door as well as the lock. Many students saw the door and lock as controls. They were able to explain how they were controls. Each student then paired off with another. The hunt had begun. The students looked at everything.

YOKASTELYN:
The chair is a control. I couldn't sit there if the chair wasn't there.

JOSE:
The bulb controls the light.

RAUL:
The fire extinguisher controls the fire.

The hunt lasted for approximately 20 minutes in the hallway. We then came back in to the classroom. Students attempted to write about the controls they had found. The students became anxious when it was time to write. In reflecting on this activity, I felt that giving names of objects in the hallway could have alleviated some of the difficulty. Some students asked for spelling, while others used phonetic strategies to spell the words. We then shared our findings.

In general, this activity went well. The students understood the assignment. They went through the hallways quietly, trying not to disturb other students working. Sharing their findings proved to be valuable. They each found something different as a control. They were able to help one another in determining if the object was a control or not. They had a successful experience, which built up their self-esteem.

In terms of assessment, my first concern was to see if they could find control devices in the hallway. Once they found one, could they explain how it controlled? I felt that the next step would be to identify the control device with the thing controlled. The day before the activity, I let the class know that a friend was coming to take pictures of them. They seemed to be very excited.

The big day came. I placed several different objects on the rug We reviewed what controls were. We began to focus on a stapler. What is it? What does it do? How does it work? What makes it move? Students began to notice things:

JOAN:
The stapler goes up and down. A screw makes it go up and down.

ANAYENCI:
The plastic strip pushes the staples.

JENNIFER:
Pega papel (attaches paper).

4-7: **Explaining an adjustable wrench**

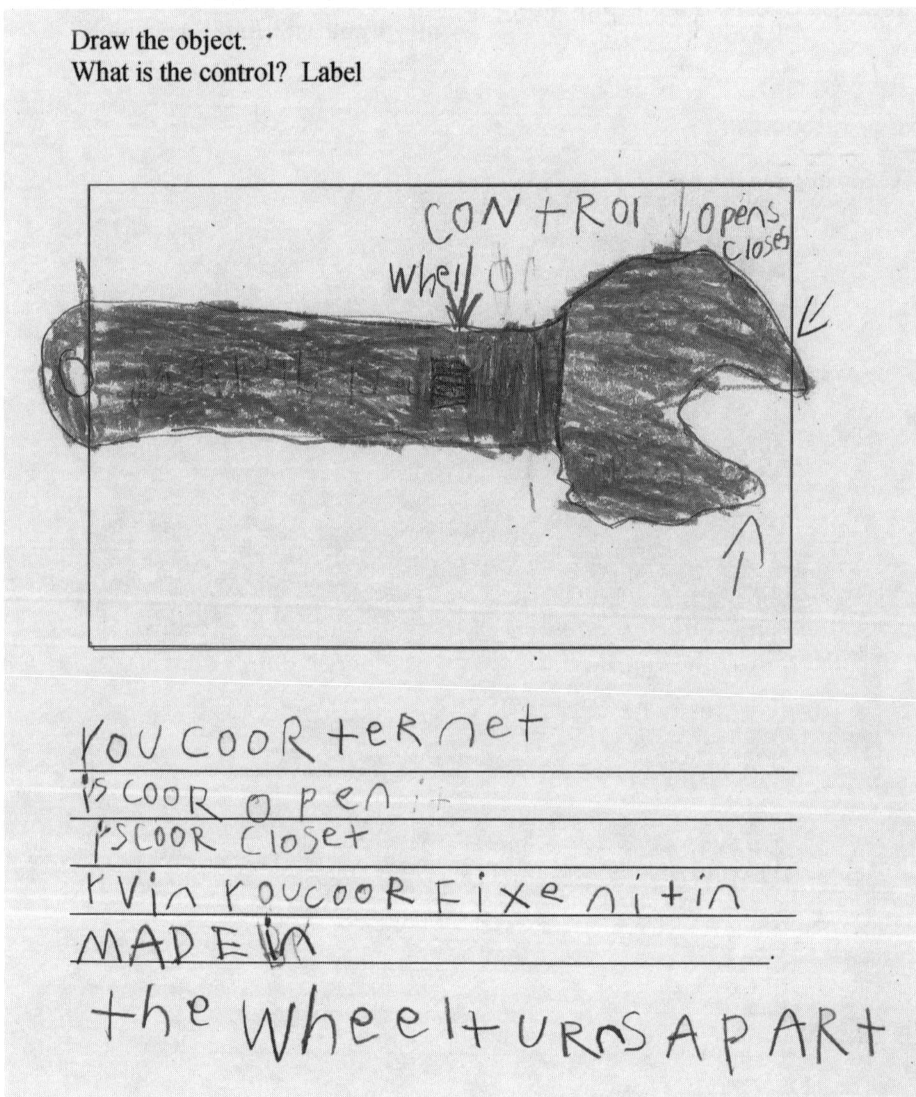

Students were then divided into pairs. Each pair was given an object. They were to draw the object, and then describe where the control item was and how it operated. The control item may have been a button that turned something on/off when pressed. It could also have been a knob that controlled the volume or made something go on/off when turned.

Most pairs worked well. They quickly set out to draw the object given. The difficulties became evident once again with the writing aspect. They seemed to lack the vocabulary necessary to describe the motions they were observing. As I went from pair to pair, I noticed that all the students were having this problem. Many students dictated their responses to me and I wrote them down in order to relieve their frustrations.

In reflecting upon this activity, I felt that I should have anticipated some of the vocabulary words that might come up, for example, "spring," "knob," "turns," "up/down," "back/forth," etc. I should have had these words available in Spanish as well as English. This might have alleviated some of their frustrations.

In terms of assessment, students were able to point out where the control was. They were also able to determine how the object worked. Their frustration was with reading and writing.

After a discussion of this problem with *Stuff That Works!* staff, Kathy decided she would write a single word on the board, such as "control." Having drawn the object and identified the control device, all the students would have to do initially was copy the word "control" from the board onto their drawing. She could later expand their written vocabulary to include more words.

The next day I tried this modification to the previous activity I had done. The only word they needed to write was "control." At this point, they were not frustrated. The activity went very smoothly. I do not know if it was because this was the second attempt to do the activity, or because of the limited writing required.

Before long, Kathy's students were writing extensively about controls. An example is shown in Figure 4-7. At the end of the school year, Kathy looked at these drawings again. She observed, "It's hard to believe that these are the same children that I have now. They have come a long way."

Second-Graders Learn About Levers

by Shirley Monterroso-Nieves

Shirley Monterroso-Nieves teaches learning-disabled second graders at P.S. 96 in East Harlem, New York City. She became intrigued with mechanisms and felt that an extended unit on levers would create important learning opportunities for her students.

I must say that initially, the topic of mechanisms did little to motivate me, and thereby enable me to teach it to my second graders. In fact, I was very intimidated by the topic although I had engaged in some exploratory learning over the summer. There were many things that I did not fully understand, and I decided to do some more reading to feel comfortable with it myself.

As I began to sift through some of the very basic concepts in mechanisms, I decided to narrow my objectives for my students. I didn't think it would be developmentally appropriate for seven- and eight-year-olds to grasp the concepts of first-, second-, and third-class levers, nor did I view it to be so important. What I did see as valid learning concepts were:

1. **Its function or job: What does the mechanism do?**
2. **Cause-and-effect: How does it work?**
3. **The role of our body in using it: What must you do to it to make it work?**

Eventually, I wanted my students to be able to identify the mechanism's parts in terms of "effort," "load," and "fulcrum." I thought that if my students could pick up any mechanism and identify its three major parts, I would have accomplished a tremendous task.

What Does It Do?

Our journey into mechanisms began with an exploratory look at how a clothespin works. First I read them a short book called *How Does it Work?* This book gives a very simple, pictorial explanation of the operation of a train locomotive. I asked, "What do you think the book was trying to show us?" Esther raised her hand and said that it was showing how different machines work.

I held up a clothespin, the kind with a spring inside, and squeezed it open. I asked if anyone knew what it was. Robert C. raised his hand and said, "It's for clothes, to hang them outside." I said, "Yes, it is indeed a clothespin," and asked if anyone knew what its job was. The room was quiet for a moment, so I prompted on. "Does it wash the clothes for us? Does it dry them or make them look pretty? What does it do?"

Kathy raised her hand and simultaneously called out, "It holds the clothing so it doesn't fall down!" I praised her and asked who else had seen or used these before, and many of the children raised their hands.

Then I wrote the word "mechanism" on the board and asked if anyone knew the word. Cathy said something like "MEE-CHEW-NISMS" and Chris yelled out "Machines!" I was elated that they had gotten that far in decoding. Also, Chris's connection with the word "machine" was a perfect lead-in to defining the word.

I asked if anyone wanted to take a guess as to what the word could possibly mean, and it became quiet again. I noted that someone had already mentioned it, and after a few seconds said, "Chris, you used the word 'machine.' That's exactly what a mechanism is, except that it is a very small, simple machine. It's not as big as a car or boat, but I'll bet there are lots of little mechanisms that help cars and boats work."

I explained that as a first step towards understanding how mechanisms work, I wanted them to try to figure out how this particular mechanism, the clothespin, works. I gave each child a clothespin and asked them to make maps of these objects. I used the term "map" instead of "diagram" since "map" was already a familiar term to them. I was not sure they were ready for terms like "fulcrum," "effort" and "load." Instead, I asked them to give the mechanism parts names like "top" and "bottom," and to identify and label the part of the mechanism that did not move. I also asked them to write a few sentences about how they thought it worked, and what you had to do to make it work.

Each of them was able to trace the outline with the help of a friend, and one or two attempted to make a sketch of what they saw. Although many of them were very adept at articulating what they had learned, very few were able to express it in writing. Except for Esther, Catherine, Pauline and Kathy, they all avoided the writing part of this activity like the plague. Figure 4-8 shows an example of the children's work.

To assist them in this process, I wrote the words "top," "bottom," and "doesn't move" on the board. Still, they did nothing to adhere to my request, and I didn't want to push the issue. I thought it might be too much for them to take in, in one lesson. I was amazed that they had an interest and were motivated at all! I guess that touching and playing with the mechanism is what appealed to them. This initial experience made me feel more at ease with the topic, and gave me ideas about how I would approach it in the future.

Shirley had an experience that was similar to Kathy's: the children loved playing with their mechanisms and drawing them, but stopped short when it came time to write. She also applied a similar strategy: writing useful words

4-8: **"Map" of a clothespin**

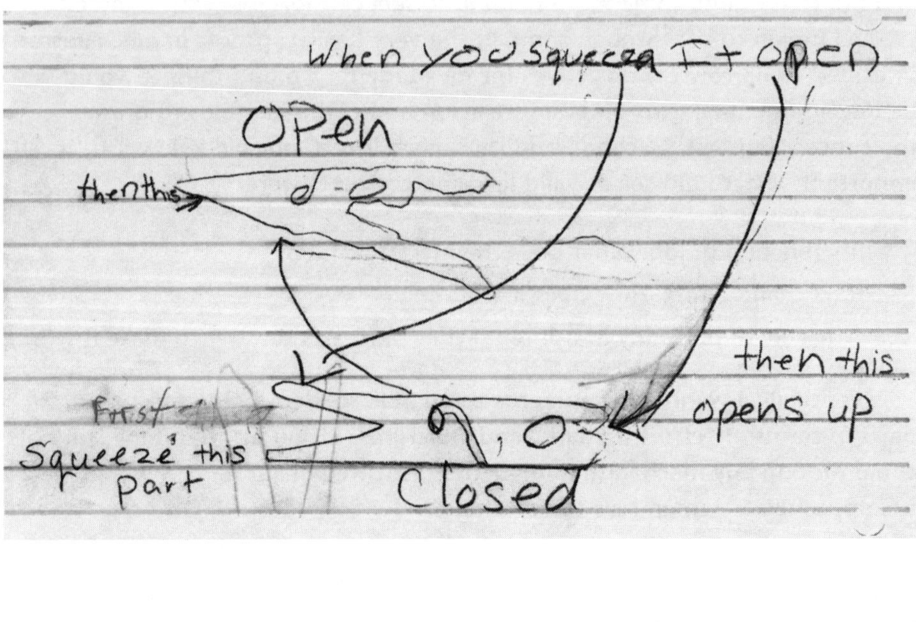

on the board for the children to copy. Although many of them did not accept the prompting, they subsequently became more confident about writing, as we shall see. *Stuff That Works!* staff asked Shirley why she had used the geometric categories "top" and "bottom" for identifying the parts of the clothespin, rather than functional categories, such as "input" and "output."

When I thought of this activity, I asked them to name the parts of the clothespin, such as top/bottom, because I wanted them to start thinking in terms of the relationship of parts to a whole. I wanted them to name the parts themselves, and see what they could come up with to establish a comfort zone. I came up with the top/bottom categories, but some of them used their own names.

My categories could have been a little fuzzy, because "top" for one child could have been "bottom" to someone else, and vice versa.

Perhaps it would have been better to label the "part that you squeeze" and the "part that does the job" – i.e., the teeth that hold the clothes on the line. I also thought about using words like "input/output," but I wanted to move eventually to terminology like "fulcrum," "effort," and "load." I didn't know whether yet another set of unfamiliar terms would just have made it more confusing. I could see them coming to me and saying, "But Ms. Nieves, you said we should call this 'input/output.' How come we have to use these new names now?"

Shirley followed the clothespin activity with a more extended challenge, which asked the children to determine the function and method of operation of a variety of mechanisms.

4-9: **One group's "squeeze" category**

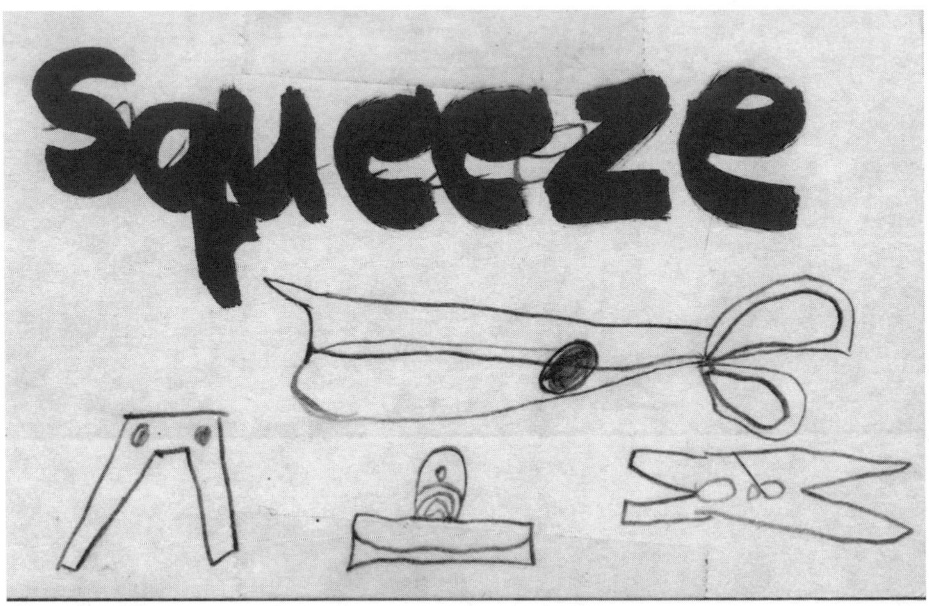

How Does It Do It?

The second activity required children to work in cooperative groups. Various mechanisms were placed on a tray, along with a large sheet of construction paper, pencils, and magic markers. I brought in the following mechanisms for them to look at:

- **Clothespins;**
- **Scissors: children's, adults', kitchen types;**
- **Spring clips: magnetic clips for a refrigerator, hair clips, bull dog clips;**
- **Ice cream scoops of different types;**
- **A corkscrew;**
- **Spatulas and large spoons;**
- **Nutcrackers;**
- **Potato peelers of different shapes and sizes;**
- **A garlic press;**
- **Can openers of different types, including a "church key";**
- **A hole punch;**
- **A screwdriver;**
- **A compass; and**
- **A door stopper.**

I asked them to do three things with these devices:

1. **Figure out what they are.**
2. **Determine what function they serve.**
3. **Categorize them according to what you have to do to make them work.**

We reviewed what a category is, and I held up various mechanisms to show what I had to do to make them work. As I was demonstrating, they came up with these categories:

- **Squeeze (Figure 4-9)**
- **Open and close**
- **Push**
- **Pull**
- **Scoop**
- **Use your hand**

The last category meant that your arm or hand becomes part of the mechanism.

Coming up with the categories was the easy part. It was another matter to get the entire group to agree on what category a particular mechanism belonged to. It was a challenge to decide who would lead the group, who would be the coach, and who would draw which object.

Miriam decided that she would solve these problems by doing it all herself. Her group was slightly upset with her, but they amused themselves anyway by further exploring the mechanisms. Meanwhile, Miriam explained in writing what she understood about the mechanisms (Figure 4-10). She was the only one to use words to categorize the mechanisms. The others used outline drawings to show which mechanisms went in each category as shown in Figure 4-9.

Afterwards, each group presented its work in front of the whole class. They had to justify why they placed the devices under each category. Most of them took this as an opportunity to pick up the objects and demonstrate how they thought they worked. This was good for the children who until then had taken a back seat in the project. It gave them an opportunity to participate orally, which some learning disabled kids are more adept at. They all had a blast!

It was nice to see them take such an interest, and it motivated me further. I would add that many of the mechanisms, which I brought from my home, were foreign objects to the children. Quite a bit of the period was taken up with trying to figure out what the mechanisms were and what job they do. They were able to identify 90% of them correctly.

Shirley's goal was to develop an understanding of levers among her children. To do this, they would need to be able to identify the fulcrum, effort, and load of some devices. She felt that the fulcrum was the easiest of the three to find, because it is the only part of a lever that does not normally move. She developed the following activity to help her children find the fulcrum of a pair of scissors.

Which Parts Move, and Which Parts Do Not?

I asked the children to take a careful look at scissors. They were to figure out what parts move, and which parts stand still, when the scissors is cutting paper. In order to show this understanding, I asked them to draw the outline of the scissors while it was closed, and another outline while it was open. Eric's work is shown in Figure 4-11. These drawings helped them see the relation of parts to the whole, and how they are interconnected. I had also brought them closer to understanding and identifying fulcrum, effort, and load, without the intimidating vocabulary that I would introduce later.

4-10: **Miriam's categories**

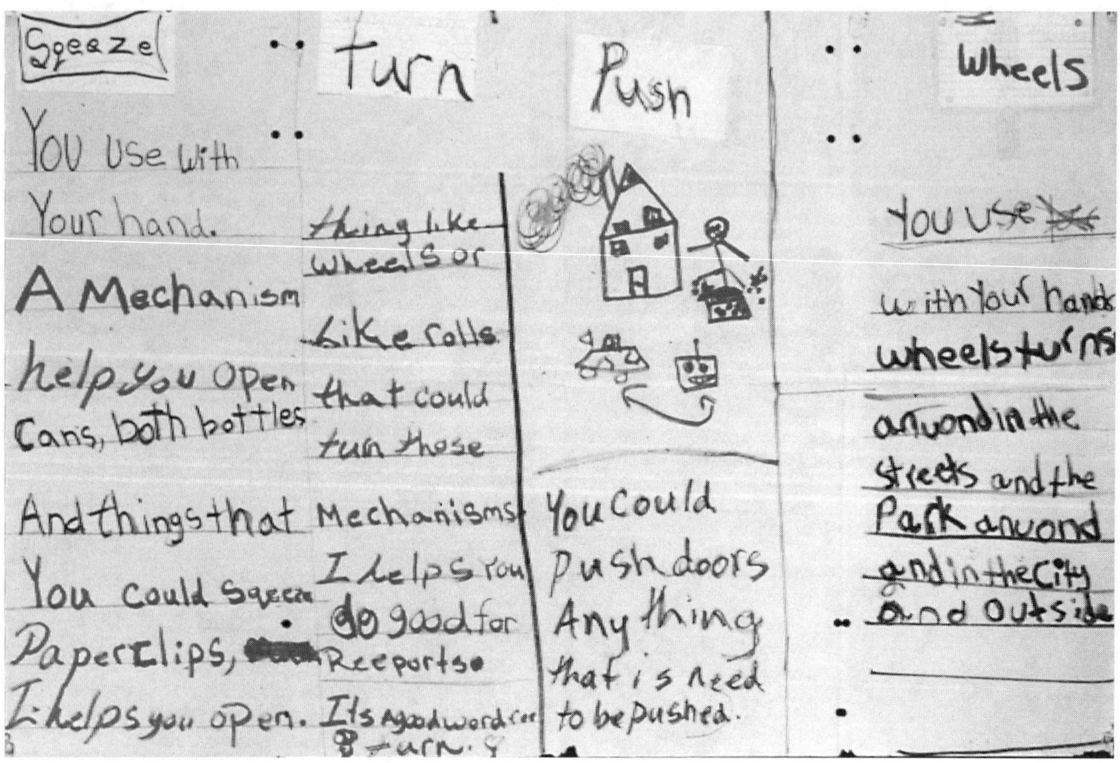

4-11: **Eric's diagrams of scissors**

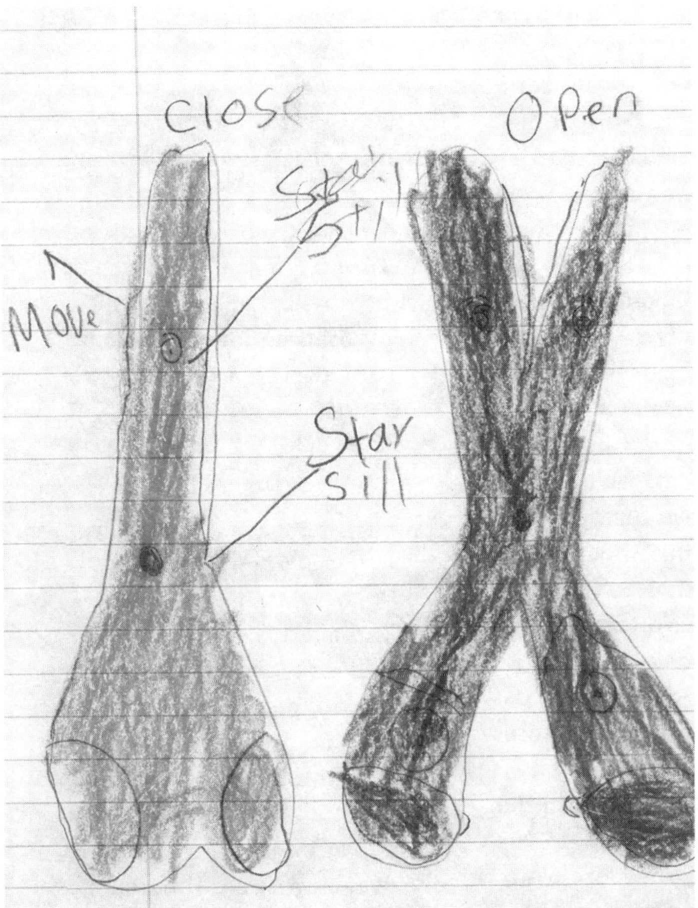

4-12: **Model of a pair of scissors**

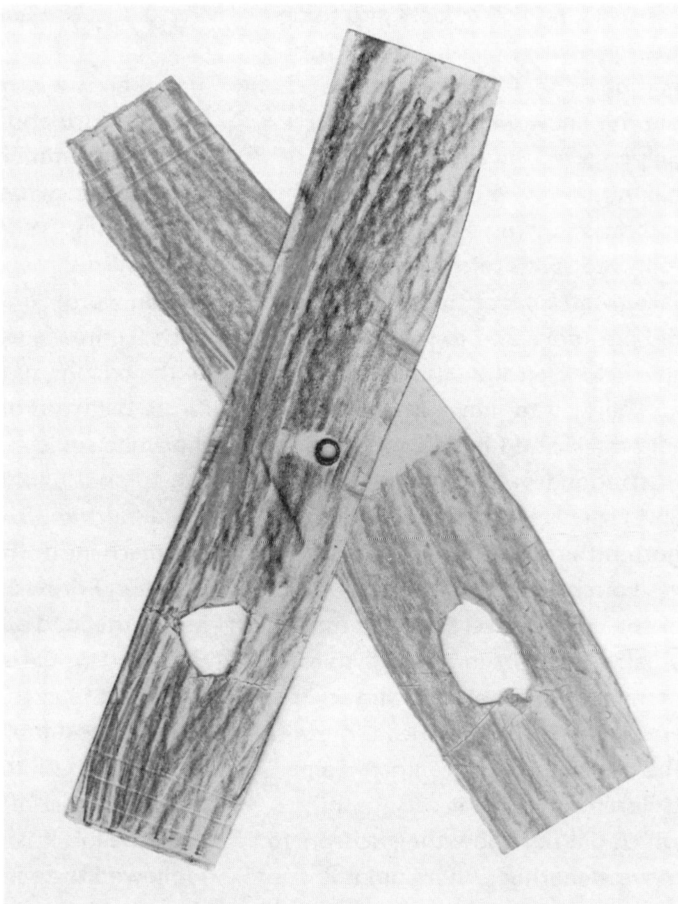

Although I had written the words "move" and "stay still" on the board to assist those who were less adept at writing, it didn't seem to help everyone. Only half the students complied with this aspect of the project. The others either didn't have time to finish or were more concerned with how the scissors looked than with how it worked.

Next, Shirley moved on to the concepts of effort ("If I do this...") and load ("...then this happens."). This approach to understanding "effort" and "load" ties these terms in with the most basic concepts of cause and effect.

Based on the dialogue that took place, I do know that approximately 95% of the class understood which parts moved and which didn't, even if they didn't write it down.
I tried to circulate around the room so that I discussed these elements with each child at least twice. If it was unclear to a child, I demonstrated what I meant by pointing to each part and asking which ones moved and which did not. By breaking the mechanism down to its parts, they were able to see the connection of the effort and the load: if I do this (effort), then this happens (load). They could also see the stationary role of the fulcrum, which holds the pieces together while allowing the flexibility of movement. It was a learning experience for all of us.

Shirley's next activity engaged the children in making models of scissors, as Annette did. Shirley used eyelets instead of paper fasteners. (Chapter 1 describes the eyelets and eyelet tool in greater detail.) Also, Shirley used the scissors models to further develop notions about fulcrum, effort, and load.

Constructing a Model

The next part required the children to construct a pair of scissors using oak tag and eyelets. Each child received a 6" x 11" piece of oak tag, a pair of scissors, and a hole punch. Five tables shared four eyelet punches. I demonstrated how the eyelet tool worked and gave them scrap sheets to practice on. I explained that the eyelet punch is more effective if you first punch a hole in each piece of oak tag with the hole punch.

The children were very excited about constructing the scissors and using the eyelet mechanism, which they had never seen before. Many of them did not know where to begin, so I helped a few by cutting the oak tag in two pieces. The rest of the class saw this as a good idea and followed suit. I told them all to take a careful look at the part that did not move and to put the eyelet there, to keep the scissors together and allow it to move. My instincts told me that if I hadn't said this, frustration levels would have risen and far fewer children would have gotten the point of the lesson.

I am happy to report that most of the children were able to construct a pair of scissors in a reasonable manner. (An example is shown in Figure 4-12.) The glitches came with the use of the eyelet punch. Some children squeezed it very tightly, which did not allow their scissors to move smoothly. Others did not squeeze tightly enough, and this led to the scissors eventually falling apart. Yet most of them were successful. I complimented them on all their hard work and ingenuity.

Shirley felt that the time had come to introduce the terms "fulcrum," "effort," and "load." She wanted to make these "science" words come alive for them, so they could also make connections with other kinds of mechanisms.

Vocabulary Build-up: Fulcrum, Effort and Load

First, I drew a large pair of scissors on the board and utilized one of the scissors the students had made, to demonstrate what I wanted them to know. I said, "When we talked about mechanisms in science, we used certain words to talk about different parts." I drew a large arrow pointing to the fulcrum of the scissors, and wrote the word underneath it. I asked if anyone could read the word. Anita called it out. I said, "Yes! We use the word 'fulcrum' to mean the part of the mechanism that does not move."

Then I drew another arrow pointing to the load and wrote the word underneath. Catherine raised her hand and read the word to the class. I said, "Good job, Catherine. We use the word 'load' to tell people that this is the part of the mechanism that does the job it is supposed to do." I followed the same procedure with the word 'effort' and Esther read it for us this time. I said, "Terrific job, Esther. We use the word 'effort' to show that we must do something to this part of the mechanism in order to get the rest of it to do its job."

Shirley then reinforced the word "effort" by using the example of "putting a lot of effort into your work." To make these ideas and terms come alive she then lay on the floor and pretended to be a pair of scissors.

I lay on the floor and asked, "Michael, will you put the effort into the scissors to make it work? And Marie, you hold the piece of paper at the other end of the scissors. Now Mike, open and close the scissors with your effort while I cut the load."

I demonstrated and the whole class began to laugh and raise their hands to get a try at it.

I asked one last question: "If my body is a pair of scissors, my feet being the effort, my arms being the load, where is the fulcrum?" I demonstrated again. They all yelled out, "Your stomach!" I said, "Yes! But why is that?" It was quiet and Esther raised her hand. "Because that's the part that doesn't move."

Shirley felt that this was a very successful unit. In her final reflections, she suggested that she had gained from it as well as her children:

All in all, this curriculum was a tremendous learning experience for all involved, particularly me. Science, for me, is an area that I am deeply attracted to but for which I feel most inadequate. I know myself as a learner, and I have a tendency to shy away from areas that seem overwhelming and intimidating. I must begin my journey as a child, one step at a time. I'm only happy to have my kids around to accompany me along the way. They make the trip much more exciting and far more rewarding.

What Shirley accomplished with these second-graders is impressive, particularly given her initial hesitation. It would be interesting to know to what extent the children's understanding of "effort," "fulcrum," and "load" would transfer to other devices. The simplest transfer would be to a similar mechanism, such as a pair of pliers. More difficult would be finding the effort, fulcrum, and load of a second- or third-class lever, such as a nutcracker or a tweezer.

Mechanisms in a Special Education Resource Room: From Pop-Up Books to Conga Machines *by Mary Flores*

Mary Flores is a resource room teacher in C.E.S. 42 in the South Bronx, the same school where Annette Purnell teaches and where Kathy Aguiar taught. She works with small groups of children, no more than eight at a time, from both lower and upper elementary grade levels.

My resource room setting is out of the norm. I have five groups with no more than eight students at a time. If my lesson doesn't work with one group, I have the opportunity to change the activity and do it differently with the next group. By the time I have finished with the fifth group, I have polished the lesson.

Mary believes very strongly in active learning. For several years, she has been using mechanisms as a topic that engages her special education students and provides opportunities for developing their literacy. Before beginning a formal unit on mechanisms, her children had been trying to make pop-ups as part of a bookmaking activity.

We began studying mechanisms back in November. However, it was not defined as such. I brought in samples of books I'd made at a book arts workshop. I included many different design possibilities for students to look at and replicate. My books included pop-up mechanisms, sliding mechanisms, and doors within the book that opened and closed.

The bookmaking served as an incentive to get the students to write. I decided to have the students examine and make the mechanisms in pop-up books. The students incorporated the designs found in a variety of pop-up books to create their own. Many of them had difficulties in this area. Some of their books took on the appearance of my models.

Several months later, Mary returned to the topic of mechanisms, with an extensive unit that made the progression from "Getting Started" through "Analysis" to "Design."

March 9

Today we began the study of mechanisms. I asked the students in one group, mainly third graders, to sit in a circle around a chart tablet. I asked them to define the word "mechanism." They were unable to, but instead began identifying objects they thought were mechanisms. William began the KWL chart by stating that mechanisms are "in vehicles and electric things." No one else responded.

I asked the students what word they could find inside the word "mechanism." Lismarie said that it sounded like the word "mechanical." I asked for an example, and she said, "like a pedal on a bike. It does something." William said, "I see the word 'mechanic.'" Other students raised their hands and I wrote all the answers on the chart tablet:

- **For vehicles/wheels on the car**
- **Electric things**
- **Pedal on a bike/it does something**
- **How stuff happens/like if you're making computers, you have to go through steps**
- **Math/symbols**
- **Calculators/pencil sharpeners**
- **Electric stuff**
- **Telephone/television/CD player**
- **Pen**
- **Things that run on batteries**

At some point, the students ran out of ideas. I said, "I am going to give you a clue about what a mechanism is. Your arm is a mechanism. Now define the term 'mechanism' for me." They added these to the list:

- **Things that move/your arm**
- **Magnet**
- **Hammer/screwdriver/door lock**
- **Tools**

After exhausting their responses, I handed out mechanisms and a worksheet entitled "What's My Mechanism?" (Figure 4-13)

The mechanisms included:

- **Eyelash curler (Figure 4-14)**
- **Garlic press**
- **Bug catcher**
- **Wire stripper**
- **Wrench**
- **Pincers (reverse tweezers)**

Many of the students did not know what their mechanisms were. Although things went relatively smoothly, I am not sure where I am going next. Do I introduce more simple mechanisms, or have them build their own?

In the afternoon, I conducted this same activity with a fourth-grade group. They were the only group that already knew what the word "mechanism" meant. I had to limit the discussion because I was anxious to get to the activity. We could

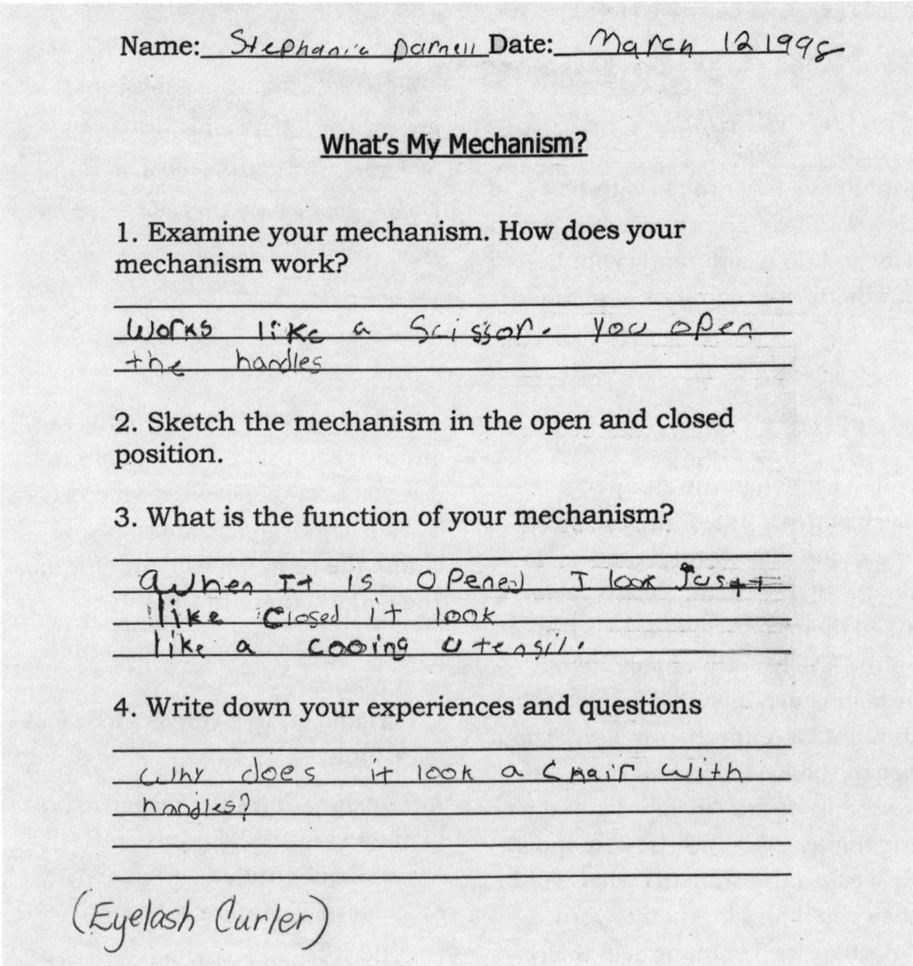
Name: Stephanie Darnell Date: March 12 1998

What's My Mechanism?

1. Examine your mechanism. How does your mechanism work?

Works like a Scissor. You open the handles

2. Sketch the mechanism in the open and closed position.

3. What is the function of your mechanism?

When It is Opened I look Just like Closed it look like a cooing utensil.

4. Write down your experiences and questions

Why does it look a Chair with handles?

(Eyelash Curler)

4-13: **"What's My Mechanism?" worksheet**

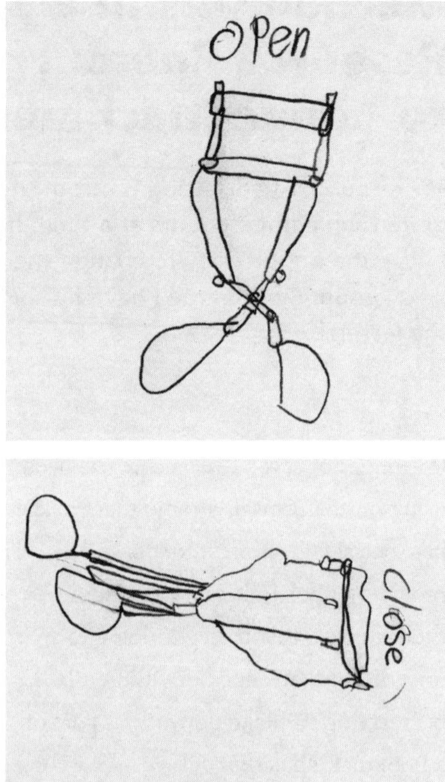

4-14: **An eyelash curler, open and closed**

have kept on going. Here is the list they generated:

- Certain kinds of tools
- Mechanical objects
- Inventions/like a robot
- Tools that do certain jobs/ monkey wrench, crank, slingshot
- Chairs: wheelchair because it has a lever, folding chair
- Ball bearing/things that move
- Computer, typewriter, fan, faucet, car, broom, door lock, door knob, staple gun, hole puncher
- "We are mechanisms because we move around." (Victoria)

What do I do with all this great stuff!

After being assigned their mechanisms, this group of students had little difficulty determining the functions of their mechanisms, as shown in Figure 4-15. I just had an idea! I will have them categorize the mechanisms.

March 18

I instructed the morning students to begin grouping the mechanisms. Things did not go smoothly. They did not understand the term "grouping." I had to think quickly. How could I get them to understand the concept? I tried to explain how to group without grouping for them.

Dario and Tanya completed the activity rather quickly. They said, "One group opens and closes and the other group does not." It was not the kind of answer I was looking for. I had wanted them to group by function. No dice!

Later that day, I conducted the same activity with my fourth-graders. Considering the problems the other groups had had with the term "grouping," I played a game with them. I asked all the students to stand in a circle, and instructed them to group themselves according to their own similarities and differences. The possible categories were endless. I wrote their ideas on the board:

- Buttons/no buttons
- Earrings/no earrings
- Blue pants/gray pants
- Keys/no keys
- Glasses/no glasses
- Teacher/students

4-15: **The eyelash curler explained**

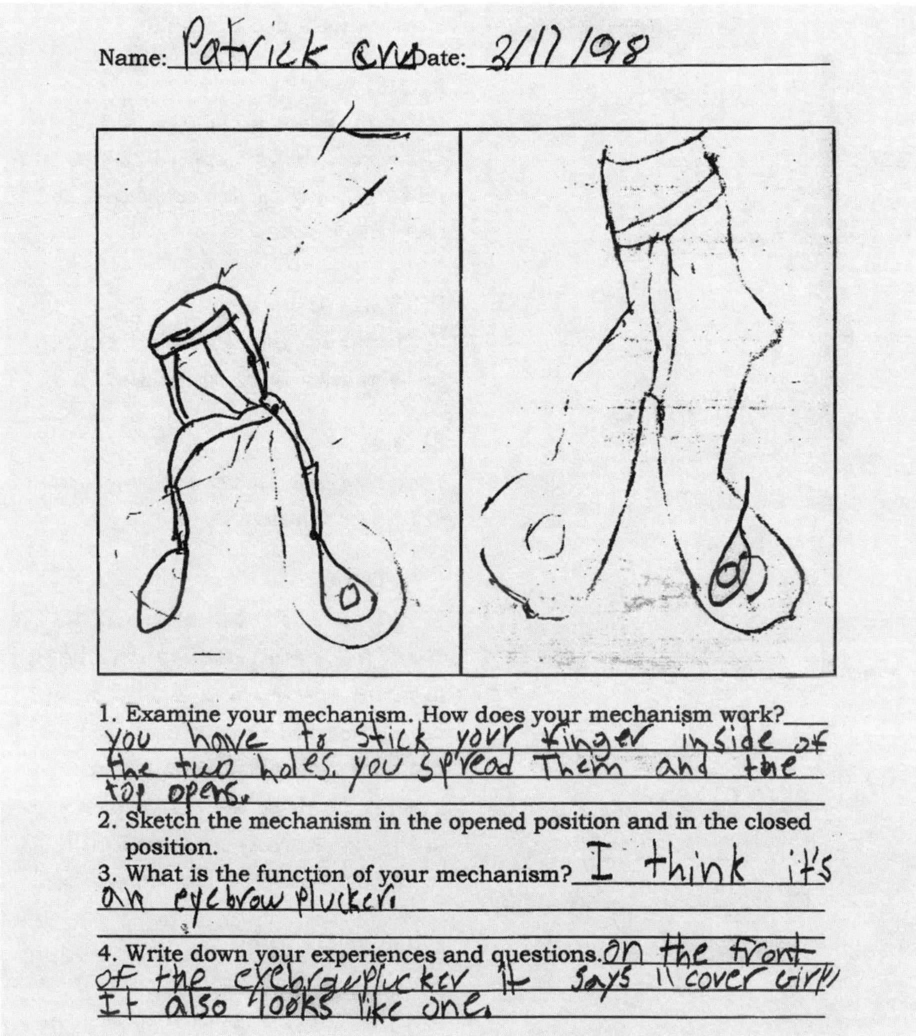
Name: Patrick Date: 3/17/98

1. Examine your mechanism. How does your mechanism work? you have to stick your finger inside of the two holes. you spread them and the top opens.
2. Sketch the mechanism in the opened position and in the closed position.
3. What is the function of your mechanism? I think it's an eyebrow plucker.
4. Write down your experiences and questions. on the front of the eyebrowplucker it says "cover girl" It also looks like one.

- Girls/boys
- Bald-headed/hair
- Barrettes/no barrettes
- Shoes/sneakers
- Jackets/no jackets
- Chains/no chains
- Belts/no belts

Afterwards, I brought over my own collection of mechanisms and had the students identify them. As they named them, I listed their responses on the dry eraser board. I identified the ones they could not name. The mechanisms were:

- Ratchet wrench
- Eyelash curler
- Retaining ring remover
- Garlic press
- Combination wrench (box/open end)
- Phillips-head screwdriver
- Adjustable wrench
- Wire stripper
- Needle nose pliers
- Scissors
- Hammer
- Pincers
- Staple remover

I divided the students into two groups and assigned each group to an area. The two areas had been set up to include similar mechanisms. Each child was handed a worksheet entitled "Where Do I Go?" (Figure 4-16)

The worksheet originally read, "Categorize the mechanisms into two groups. Tell why you divided them the way you did." I realized that this task was too simple for this group. I went over to the computer and changed the worksheet to read "Categorize the mechanisms. List all the ways you can divide them."

Moises and Cynthia worked cooperatively. They took turns categorizing the mechanisms as Moises wrote down his group's responses. At one point, Moises called me over and said, "Can you figure out how we categorized them?" He had turned the tables on me!

Derrell and Victoria were having trouble working cooperatively. They were each trying to categorize the mechanisms individually, instead of working together. I told them that they had to resolve the problem in order to complete the task. I noticed that Derrell had found six ways to categorize the mechanisms and Victoria had found five (Figure 4-17). I informed them that if they put their responses together, they would have a total of 11 categories, which was more than the other teams had! They liked that idea, and completed the assignment without further incident.

Mary felt that she had accomplished a great deal with this last group. The grouping game was a natural lead-in to the categorizing activity, and the students had come up with some exciting discoveries. But what should she do next? Mary wanted to engage her students in some design activities, but felt that they needed some preparation. She decided to introduce the concept of simple machines. Analyzing the operation of these basic devices would give them the background they needed for design.

4-16: **"Where Do I Go?" worksheet (original version)**

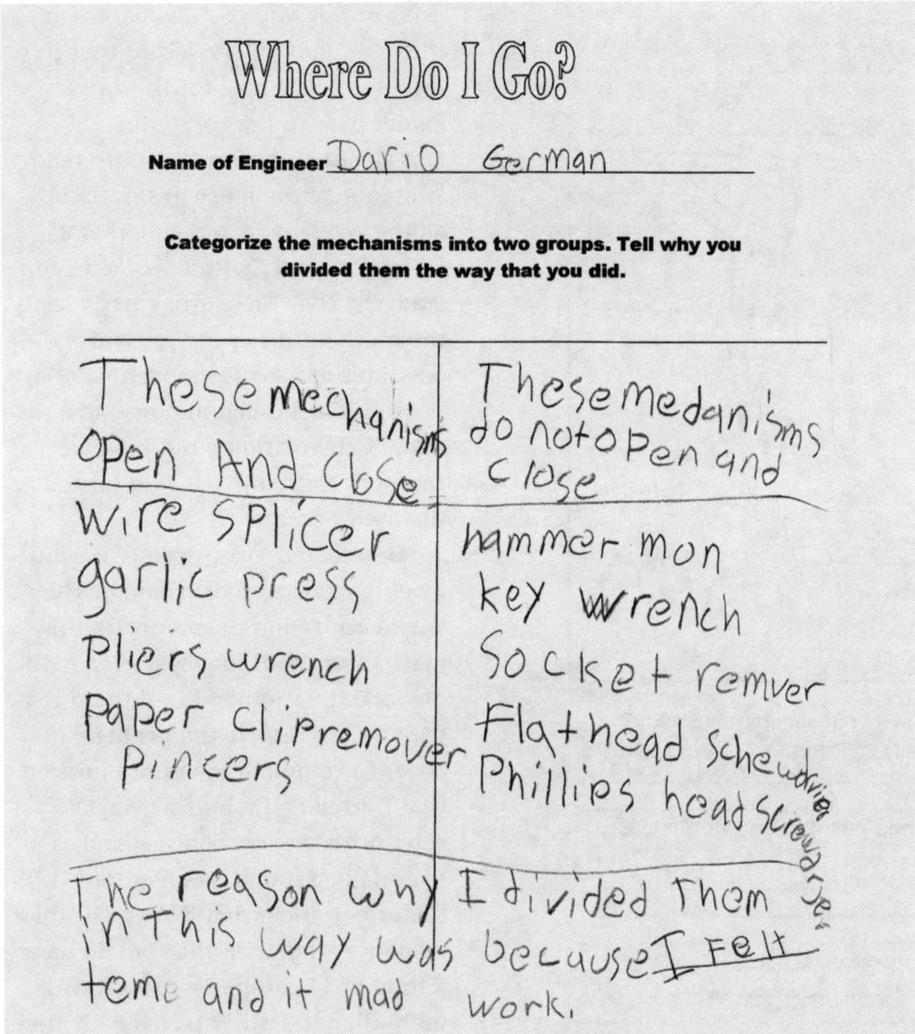

Where Do I Go?

Name of Engineer Dario Gorman

Categorize the mechanisms into two groups. Tell why you divided them the way that you did.

These mechanisms open And close	These medanisms do notopen and close
wire splicer	hammer mon
garlic press	key wrench
Pliers wrench	Socket remver
paper clipremover	Flathead Scheudriver
Pincers	Phillips head Screwdriver

The reason why I divided them in this way was because I felt teme and it mad work.

March 24

I spent the last few days planning my next move. I decided to introduce the concept that mechanisms are composed of simple machines. I placed various objects on the floor and asked the students to tell me which were the simple machines. The items included the following:

- A wheel,
- A pulley,
- A wedge,
- A lever, and
- A ruler.

The ruler, of course, was the spoiler. They fared well. Once they had found the simple machines, I picked each one up and asked, "How would this machine make your job easier?" Here are some of their responses:

MOISES:
If you have four wheels, you can make a car and it can carry the heavy loads.

DERRELL:
You can put a string on it and roll it up.

Teacher:
How would I use the pulley?

MOISES:
If I put a pulley on each side of the rope in this room, I can send a letter to my girlfriend who's on the other side.

PRISCILLA:
To hang up clothes and to put up pictures in your room...

EBONY:
You can use it to pull up water to your window.

CYNTHIA:
Say you want to get a toy that fell, and you live on the second floor. You don't want to go outside and get it, so you get that and tie a string that goes down and you take the string and you pull it up so the toy can come up.

EBONY:
You tie a fishing rod and take a long string. You pull it up and you pull a fish out of the water and you cook it.

VICTORIA:
You could, like, put a rubber band between the room and then you hang it

PRISCILLA:
When you're going to work, you could put your suitcase on the pulley and let it slide to the other side, and then you could grab it and go to work easily.

I felt that the students were making connections between the simple machines and real life experiences. My hope is that they will apply these understandings to the design of their own mechanisms.

4-17: **"Where Do I Go?" worksheet (new version)**

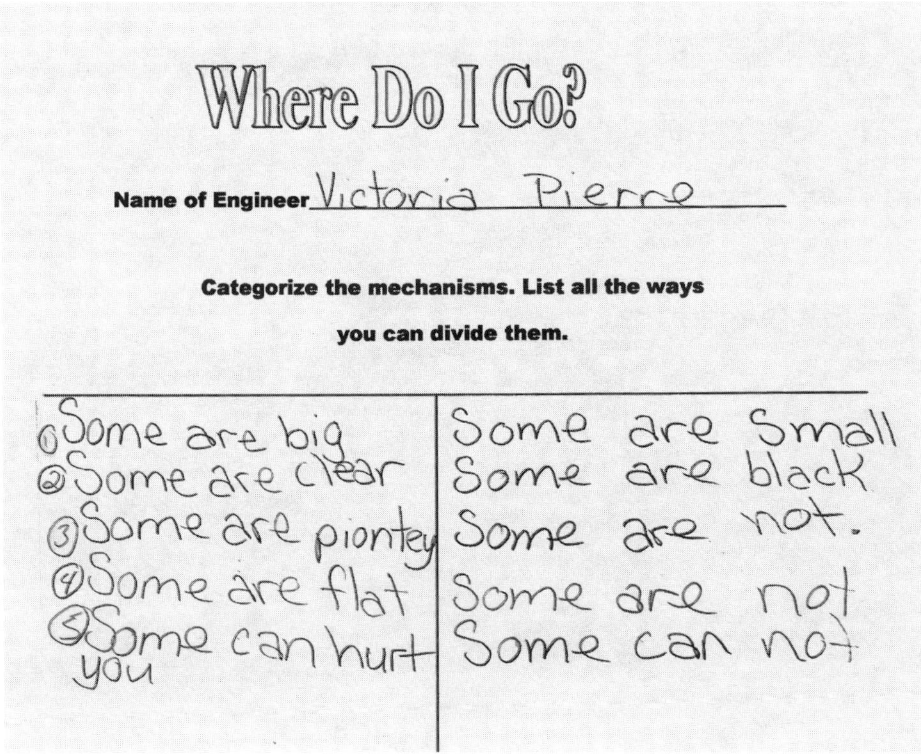

Where Do I Go?

Name of Engineer Victoria Pierre

Categorize the mechanisms. List all the ways you can divide them.

① Some are big	Some are small
② Some are clear	Some are black
③ Some are piontey	Some are not.
④ Some are flat	Some are not
⑤ Some can hurt you	Some can not

March 25

I got the idea to have the students examine some illustrations of "Rube Goldberg" devices and figure out how they work. In particular, I wanted them to find the simple machines in his wacky inventions. After discussing whether his inventions would work, I gave them the following challenge:

"Design a mechanism that would allow you to stand away from a table, and turn a page in a book on the table without touching it with your hands."

They began by mapping their designs. Moises's plan (Figure 4-18) suggests that his probably could work.

Prior knowledge of mapping helped them design blueprints that showed detail (Figure 4-19).

I tried to get them to think about some of the simple machines that could be used to make their inventions. I showed them the materials that I had for them to use: wooden spools, wheels, empty paper towel rolls, toilet paper tubes, Styrofoam, paper cups, etc.

March 30

Students in my fourth- and fifth-grade group retrieved their maps and set out to do the task at hand. Patrick, Heriberto, and Ricardo gathered their materials and began constructing their devices. They needed very little prompting from me.

Nicole's invention was not designed to open a book, but rather was a "Conga Dancing Machine" that will "dress me for Flamenco in the morning." Nicole and Shelva were the only students who strayed from the assignment. I was pleased that they were all so engaged. When I called time, they were disappointed. We still have not had time to share.

March 31

My fourth- and fifth-grade group entered the room and plunged right into their work. I moved around the room to watch and listen as the children engaged in the activity. It is the stepping back and observing that allows me to learn and understand from a child's perspective.

Patrick built a windlass wind-up mechanism. He placed a broomstick through a plastic crate and attached a ruler to the end of the broomstick. The ruler served as a lever for turning the broomstick. I asked him where he got the idea, and he said, "A book." It was similar to a design found in the book *Wheels at Work.* He felt a sense of accomplishment. After testing his invention, he stated, "I want to make another one." He found a box and began building his second windlass.

It is the reworking that allows students to move to a higher level. Patrick is demonstrating a form of self-evaluation by reflecting on his progress and redesigning his invention to make it better. Christina is using the eyelet tool to make linkages that extend outward. She attempted to use the linked pieces as a ramp. It was not rigid enough. There were too many joints. She needs to redesign her mechanism.

4-18: **Moises's plan for a page-turning machine**

Moises

I am making a mechanism that will open a book from 3 foot away. I am going to use two ramps, two balls, baster, a soda botell, pan, Wates, a hand, a Book, a desk,
Step 1. You push a big ball.
Step 2. the Ball hits a See saw on the left saed.
Step 3. the seesaw movies and the right saed up.
Step 4. on the right saed it have a Wates.
Step 5. the wates go into the ari and hit a ari blower.
Step 6. the ari blower have a ball init the Ball flys out.
Step 7. The Ball hit a soda botell. the top fly off.
Step 8. The soda go out of the botells.
Step 9. The soda hit a pan and a sponge.
Step 10. The pan go on the seesaw and The hand go up and open the Book.
Step 11. The pan falls and hit the Ball.
Step 12 the Ball rouls on the floor.
Step 13 the Ball go up a ramp.
Step 14 the Ball flatines the pg. of the Book

4-19: **Ricardo's book-opening machine**

It rarely happens that a newly designed product works perfectly the first time. Design is a back-and-forth process, where the problems with one design lead to improvements in the next. Too often, children are socialized to believe that failure is to be avoided at all cost and eradicated whenever it does occur. As a result, they find it difficult to accept failure as an inevitable and productive feature of the design process. Mary's students have reached the point where they can look at their own designs and learn from their shortcomings. They see the failures in their designs as opportunities to do better next time. This is a process that even adults have difficulties with. It is a remarkable achievement for these youngsters.

Watching the students design, test and redesign their mechanisms is encouraging. I have gained insight into how students come to make meaning. I can see that they have applied many of the concepts that they have learned. I attribute their learning, in part, to prior knowledge in the areas of mapping and environmental analysis and design. Additionally, these activities allow students to self-evaluate their progress. I usually ask students, "Can you explain what you did?" I want students to be able to explain the process, so I can be more tuned in to their learning.

April 1

Disaster struck the resource room today, and this is no April Fool's joke. My morning fourth- and fifth-grade group entered the classroom and began a futile search for their mechanisms. I asked if they remembered where they had left them. They answered, "By the chalkboard." My heart sank. The chalkboard is beside the garbage can. I suspect that the custodian mistook the mechanisms for garbage and disposed of them.

I could not hide my disappointment. However, the students took it much better than I did. Heriberto commented, "Don't worry, Ms. Flores. We'll just do it again, and this time I'll do it better." Talk about redesign!

Fortunately, the afternoon group had placed their inventions in a box, which was still intact. Cynthia, Ebony, and Victoria chose to work on Cynthia's design (Figure 4-20). I know that as they attempt to build this mechanism, they will run into many problems. They will have to learn of the flaws in the design through trial-and-error.

I invited several of the students to share the process they were going through as they tried to build their mechanisms. I thought today was going to be disastrous, but it could not have gone better! Here is some of the dialogue:

TEACHER:
Can you discuss the process you went through in building your invention?

PRISCILLA:
I began by making the hand with cardboard. Then I decorated the box. I got the string to make my project. My mechanism did not work. My hand did not work because I put too much tape, and it's breaking.

TEACHER:
How will you redesign your invention?

PRISCILLA:
I will make it work by putting a little bit of tape, and making a new hand and putting string where it belongs.

TEACHER:
How is this project making you feel?

PRISCILLA:
It's making me angry, because I tried hard to make my mechanism work, but now I have to make another one.

TEACHER:
Moises, can you explain how you approached the activity?

4-20: **Cynthia's design**

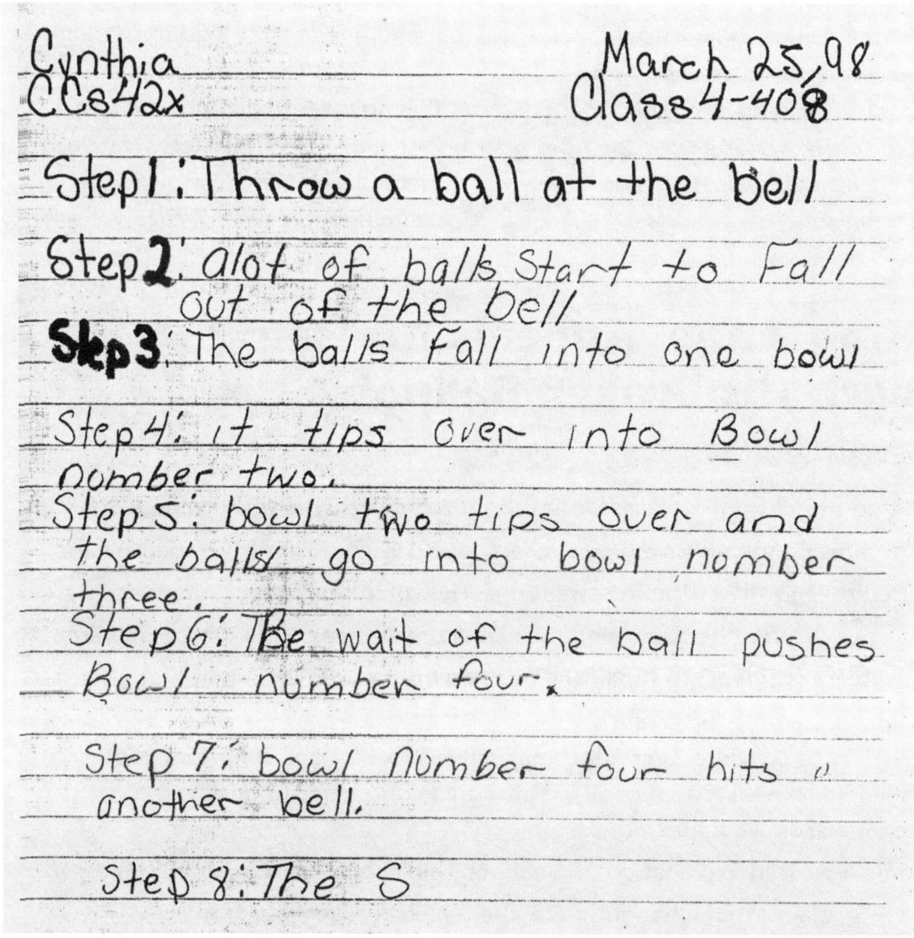
Cynthia March 25, 98
CC842x Class 4-408
Step 1: Throw a ball at the bell
Step 2: alot of balls Start to Fall out of the bell
Step 3: The balls Fall into one bowl
Step 4: it tips over into Bowl number two.
Step 5: bowl two tips over and the balls go into bowl number three.
Step 6: The wait of the ball pushes Bowl number four.
Step 7: bowl number four hits a another bell.
Step 8: The S

MOISES:
I made a blueprint. Then I wrote the process. Then I started to get all the materials. I made the ramp from cardboard. Then I made the hand. That was the hardest thing to build. That's because the first time I made it, I made the fingers out of regular pieces of paper. Every time I tried to flip the page, the fingers flipped over. Then I had to make it again. This time, I made the fingers out of cardboard. It worked. I improved the ramp, because it was always bending. I put two sheets of cardboard, one straight and the other one a little bent. It made it stronger. Then I started to look around the room to get ideas for how I can build the air pressure thing (a meat baster). I decided to use cardboard and a little bit of string. I'm going to have to get some plastic from my house.

TEACHER:
What kind of plastic?

MOISES:
You know, like, the plastic that's on the umbrella so it can stretch a little bit. I'm going to need something to keep the air in. Then I'll push it out. It has to be fast so it can push the marble inside it. I have to have something that's heavy so it can hit the baster.

TEACHER:
How does it feel to be an inventor?

MOISES:
It feels like, you know, like when you're having a test, and all of a sudden you feel that tingly feeling in your stomach, and you forget the words. That's how it feels.

I believe that it is crucial to set aside time for this kind of sharing. I know that teachers struggle with finding ways to capture children's language. What facilitates my documentation is that I can sit at the computer and type their responses as the students answer my questions. These discussions allow me to assess the student's progress. I want to know that the students can design, test, and redesign their inventions until they are satisfied.

King Angel Appeals for Help with His Broken Ballpoint Pens!

by Angel Gonzalez

Angel Gonzalez is the only science cluster teacher (specialist) in the Family Academy, a small public school in Central Harlem. He sees the school's two fifth-grade classes for two 90-minute periods each per week.

"King Angel wants to free his imprisoned subjects (the fifth-grade serfs) from his dungeon. However, vandals have damaged all of the special pens used to sign the official release forms. He would like to hire someone to fix his pens so that he could sign their freedom decrees. King Angel will hire a subject (student) to repair the pens if s/he can convince him that s/he is a capable pen-repair technician.

"Try to convince King Angel that you can fix his pens. You want your freedom, as well as that of all the fifth-grade prisoners. To qualify as a repairperson, you must provide:

- **A written explanation of what you think could be damaged in the pens;**
- **A diagram of what you think the pen looks like inside.**

You will be called to share your expertise before the King and his subjects."

After presenting this challenge, Angel showed his students one of the pens and demonstrated it, without taking it apart. Here is what Nerisa wrote:

Dear King Angel:

I'm a trained technician. I've been through science school, college and high school. Ever since I was little, I wanted to be a scientist. I fixed my little brother's toy car because it was broken. I've been fixing things ever since.

- The spring could be broken
- The bottom part could be rotten
- The ink pen could need more ink
- The clicker is probably not good, or a piece is probably broken inside.

Sincerely, Nerisa,
Your technician

Then Angel had the students work in pairs. He distributed a pen and a work tray to each pair. Putting the small parts in a tray prevents them from rolling away or getting lost. Angel emphasized that the pens must be returned in working order. Each group had to sketch the parts of the pen and explain how it works and how it might be fixed.

They enjoyed the challenge and took it up enthusiastically. (See Figure 4-21.) To illustrate the kinds of drawings I wanted, I showed them the exploded drawings of mechanisms and appliances in the book *Visual Dictionary of Everyday Things*. This book illustrates well what I wanted them to do.

These pens have an automatic release feature that retracts the pen when the clip is lifted slightly, for example, by putting the pen in a shirt pocket. Nicole's drawings are shown in Figure 4-22. What Nicole calls "the puller" is the pocket clip.

NICOLE wrote:

The spring helps the pen go up. On the pen there is oil so the spring could go up easily. The sucker is the thing that spins. It is connected to the pusher. The pusher is what you push. The top part shows the sucker and part of the ink tube. The bottom part holds the sucker and the pusher. I call the sucker the sucker because it looks like a fancy straw. I call the spring the spring because that is its name and because I could not think of another name. I called the top part that because that's what it is. The puller is what you pull to make the pen go back in. The ink holder holds the ink. Without that the ink would be all over the place. The pusher makes the pen go up so we can write. The bottom part holds the puller, pusher, and sucker. There is wax so the ink won't fall out.

Angel felt that the activity had gone well. The story about the king drew them in and generated excitement. They were able to compare their initial hypotheses with what they actually found inside. Most important, they began to develop a sense of how a collection of parts can function together to form a mechanical system that is much greater than the sum of its parts. He followed the ballpoint pen challenge with a similar activity involving water-supply valves. Again, he presented the students with a hypothetical situation:

"Water is spilling everywhere. It's being wasted. There's a drought. You need to save water and conserve. Your job is to fix the valve that's wasting millions of gallons, the valve of a garden faucet outside of a house. The town will reward the valve repair person with one million dollars."

4-21: **Esterling analyzes and sketches the pen parts.**

4-22: **Nicole's drawings of the pen parts**

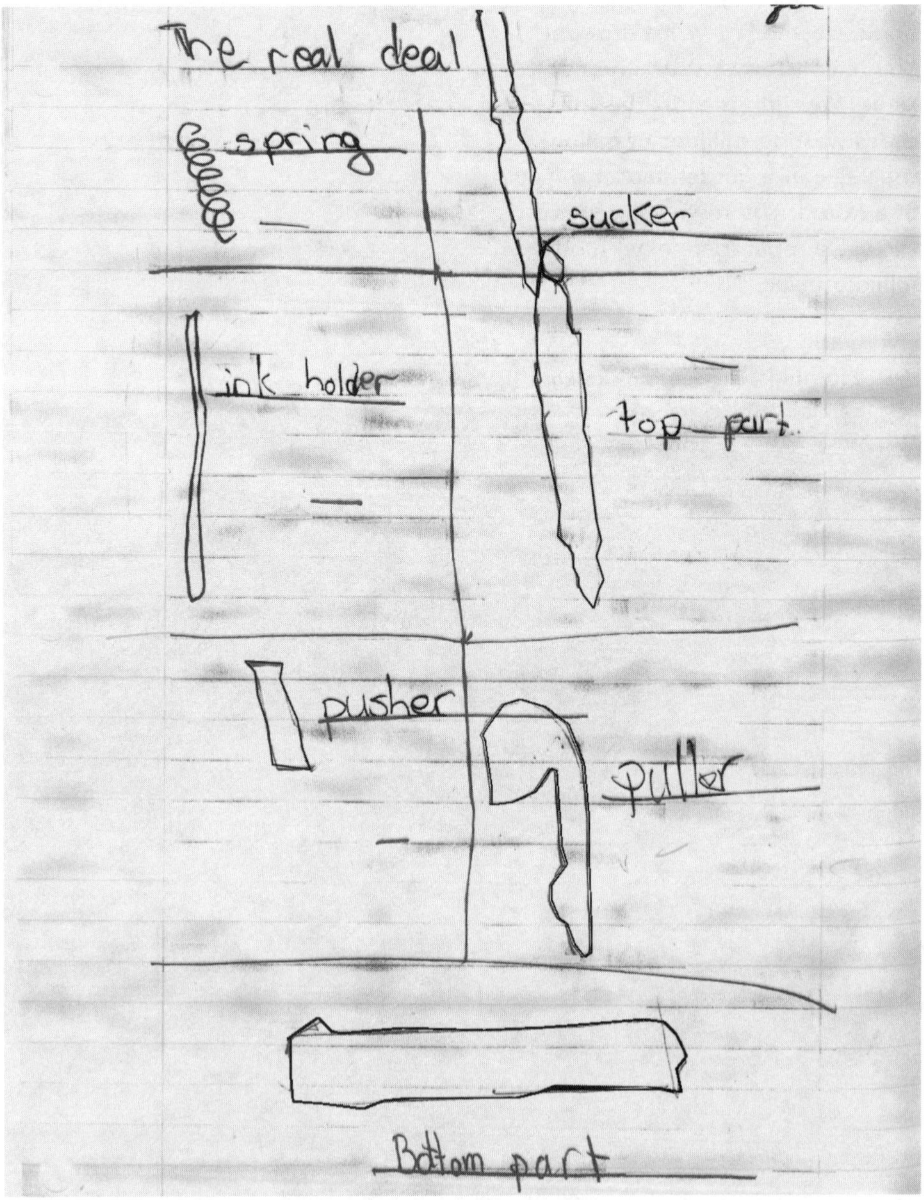

After presenting this situation, Angel led a discussion about valves and faucets. Where are they found? What do they do? How can you tell if they are not working? He then showed them an example of a valve, without letting them too near it, and asked them to draw what they thought was inside. He writes:

The students were motivated by the hypothetical situation. They were anxious to look inside the valve, but I refused to let them until they had first drawn a picture of what they thought was there. After the drawings were done, I called on Unslo and Setaire to share their ideas on the overhead projector. The overhead has proven to be a great stimulus that garners the class's attention. Setaire explained that a door had broken in the valve and was no longer controlling the flow.

Next, Angel distributed a valve to each group. After they had spent a few minutes playing with them, he provided screwdrivers so they could take them apart. He had already loosened the screws, so they would not be too difficult to remove. Esterling is shown at work in Figure 4-23. Tenisha made a drawing of the inside of the valve, shown in Figure 4-24, and wrote the following explanation:

The valve works by your turning it and a black piece comes up to let water out. When you turn it again the black piece goes down and the water stops. It was interesting because me and Dashaye replicated a valve. We learned how it works and how to build one.

Angel saw this work as a prelude to learning about switches and other electrical controls, which are discussed in the next section. He wrote the following evaluation:

Both fifth-grade classes enjoyed dissecting the valves and putting them back together. I feel that these initial experiences in hypothesizing, observing, sketching and writing about how a control works will lay the groundwork for the upcoming work of analyzing and creating their own controls. They are clearer on what a control is. I will keep the class focused on those devices that manipulate/influence something else outside of their own structures.

Angel's concluding comment refers to the confusion between the technical definition of "control" and the word's everyday meaning. He wanted the children to realize that the control must be distinct from the thing controlled.

4-23: **Esterling dissects a valve.**

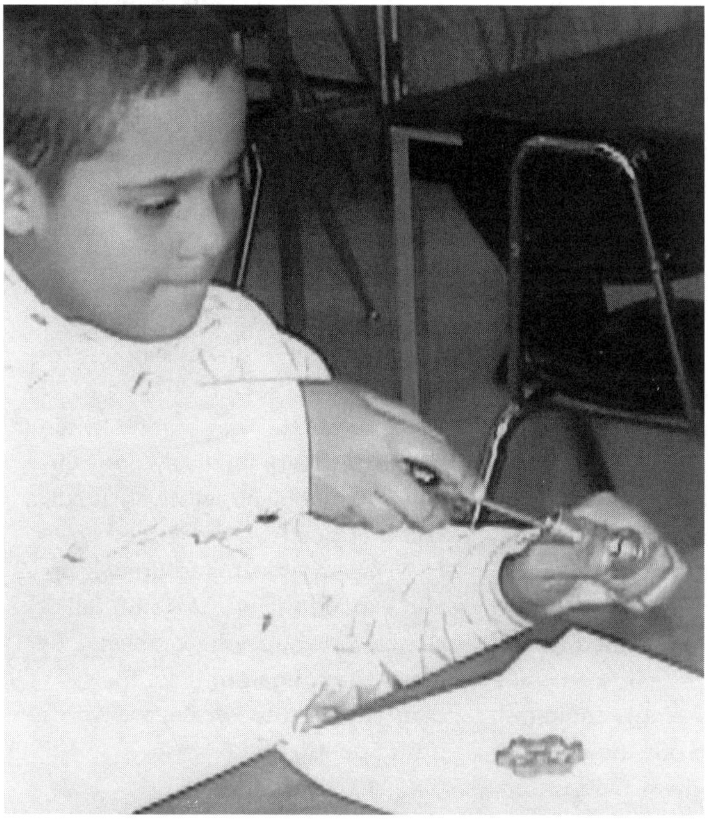

4-24: **Tenisha's sketch of the inside of a valve**

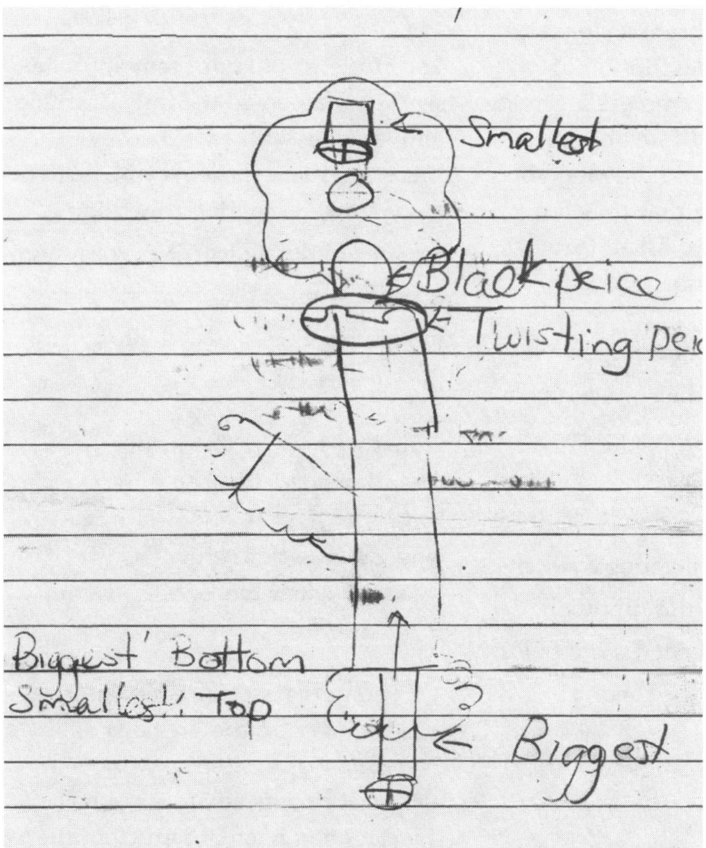

Two Teachers' Stories About Teaching Circuits

Kathy Aguiar, whose work with mechanical controls is described above, turned to electric circuits later in the year. By this point, her children had become much more confident about writing. They designed their own circuits with switches and wrote about how they worked. Angel Gonzalez, whose work with pens and valves we have just seen, describes how his students devised their own switches from common materials, and used them to create "electric code senders."

Third-Graders Teach Each Other Circuits

by Kathy Aguiar

Kathy Aguiar's unit on mechanical controls had taken place near the beginning of the school year. At that time, many of her students would become very frustrated when asked to write. Later in the year, she returned to controls, focusing this time on electrical controls.

Today I began exploring electricity with my students. Since only a handful of students came to school, I thought it would be a fun activity to do. I also thought that I would be able to give one-to-one instruction to those who might need it.

I began the lesson by asking what electricity is. The responses included:

ANDREW:
It gives you a shock.

CARLOS:
It helps to turn on the TV

RAUL:
You can plug the light or stereo on.

I then continued by asking what electricity does for us. Again the responses varied.

ANDREW:
It puts the Nintendo on.

FREDDY:
It helps to put on the light in the refrigerator.

CARLOS:
It turns things on.

We got into a discussion as to what items use electricity. Many objects were discussed. I then read excerpts from a book called *Electricity and Magnetism*. This book helped to clarify all the things we had discussed.

Next, Kathy asked her students to construct circuits using batteries, bulbs, and wires. This is the basic activity of science curriculum units on "Batteries and Bulbs," but for Kathy it was only a prelude to what would come next.

Once the book was read, I told the students that they were going to be given a battery, wires, light bulb, and a battery holder. They then would have to find a way to light the bulb. The class of eight students was divided into three groups. The groups went to work. Carlos and Natalie's group finished first. There was so much excitement when Carlos lit the bulb. Members of the other groups were very curious about how Carlos was able to light the bulb. They came immediately to Carlos's table.

At this point, I wasn't sure if I should let them look at the set-up or not. I decided to send the other students back to their own groups. I asked, "What's the fun of finding out from someone else?" I instructed them to try to light the bulb on their own. The other groups did eventually light the bulb.

Two out of the three groups had a difficult time for different reasons. One group had many leaders but no followers. Once they were able to settle upon a chief, the rest became easy. The only direction I gave to the group was that they were to decide on a way to work together. Otherwise they would not be able to finish the task on time. This seemed to motivate them.

The third group simply could not determine where to put the wires or how to connect them. This group received some help from Carlos, which enabled them to complete the task. In terms of assessment, each student drew a diagram, labeled it, and wrote a brief description of the steps taken to complete a circuit. They did not seem to be apprehensive about writing and drawing about the work. I suppose that they are accustomed to being asked for drawings and writings.

What happened next reflects the importance of providing a rich supply of materials in the classroom. It also shows how Kathy followed the students' interests as they began experimenting with switches. Incorporating a switch in the circuit moves this unit well beyond the typical "Batteries and Bulbs" activity, and revisits the concept of a "control."

Once several of the students finished their assignment, they began to look at other things I had. Some of them found a switch. I observed a group of three students trying to incorporate the switch in an existing circuit. I let them try for a few minutes before moving them on to another task. They let me know in which direction to move. My next lesson will be to have them put the switch on the circuit.

Over all, this was a good lesson in that there were relatively no problems. Working cooperatively is still a skill that needs to be worked on. Each group was able to draw a circuit and explain the workings of the circuit. Vocabulary became an issue when they began to label their diagram. I wrote vocabulary words in both Spanish and English.

- **Wires — alambres — cables**
- **Battery — bateria — pila**
- **Bulb — bombilla**

The students did not seem apprehensive. They accepted the task of writing and labeling with ease. However, there are some students who will always take exception to writing.

One of the students Kathy has been concerned about did the work shown in Figure 4-25. Translated into English, the paper reads,

"First, I put in the battery. Then I placed the red wire on the battery. Then I put the black wire on the battery. Then I put the wires on the light and it lit up!"

The next lesson (a week later) was based upon the previous one. Everyone was excited to work with electrical circuits again. Since half the class had been absent from the

4-25: **Explanation of how to light the bulb**

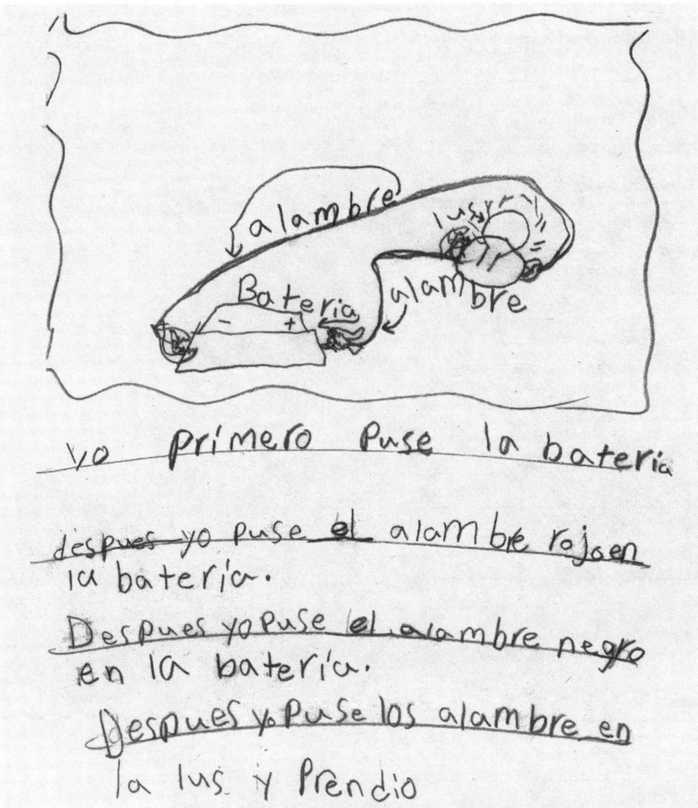

4-26: **Explanation of the circuit with the switch**

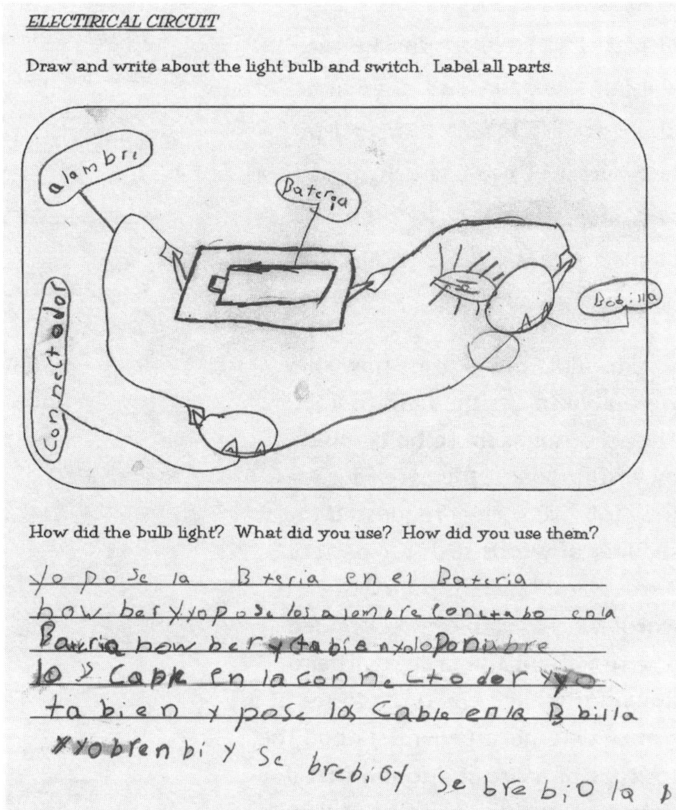

previous lesson, I decided to re-read the excerpts from *Electricity and Magnetism*. Students who had been present the first time had a better understanding of the excerpts. Those who were hearing it for the first time asked questions that the other students were able to answer.

YOKASTELYN:

What are these things?

FREDDY:

Wires.

YOKASTELYN:

What do they do?

FREDDY:

They take the current from the battery to the bulb.

This exchange (which took place in Spanish) surprised me because Freddy usually does not provide accurate answers.

Afterwards, we began to discuss the purpose of a switch. The consensus was that the switch could turn something on or off, make it loud or soft, or change a station. It could be something you would push, pull, or turn. Students were given two tasks. First, they were to build battery/bulb circuits. For the second task, students were to incorporate switches in their circuits. The switch was to put the light on or off. Students were then divided into four groups. Students who had done the previous lesson were not in the same groups as those who had not. Each group was given a battery, battery holder, three wires, and a switch.

Students were very eager to begin. They began to build the circuit without the switch. Two out of the four groups did not have a problem with the basic battery/bulb circuit. They moved on quickly to attempting to incorporate the switch into the circuit. These groups had difficulties in putting the switch in. The other two groups had difficulties with the first task. They had not been able to make the bulb go on. Some students from one group went to the other groups for help. Once they found what they were looking for, they went back to their original group. They were able to turn the light on and off using the switch.

As usual, Kathy used their work as an opportunity for them to express themselves in writing. She asked them to draw their circuits and describe the components and how they used them. An example is shown in Figure 4-26.

Translated, it reads:

"I put the battery in the battery holder. I connected the wires to the battery holder. Also I put the wires in the switch, and I put the wires on the bulb. I turned the switch on and the light came on."

Students did express how they were able to get the bulb to light. They did not seem to be troubled by writing about the procedure they followed. They requested the spelling of words such as "battery," "wire," and "switch." Once the words were given, they continued to write about the process. Their drawings really helped me to see at a glance whether they understood the concept of a circuit. I found that if a student could not draw the circuit, then more than likely that student could not explain it or write about it.

Overall, this was a good lesson. It sparked the students' excitement. Students were very motivated. They were willing to work cooperatively. The only difficulty I kept running into was with vocabulary. I try to anticipate all the necessary vocabulary words students might need in English and Spanish. However, there have been times where certain words have come up that I was not prepared for.

In terms of assessment, I knew that a student understood what they had done if the student was able to:

- **Match their drawing with their written work, and**
- **Explain orally what they had written.**

For homework, students were to find 10 items that required switches. One list is shown in Figure 4-27.

4-27: **List of ten things controlled by a switch**

A. Write the names of 10 things at home that have a switch.

1. the tv
2. the reFrigeore
3. the wash
4. the toy baker
5. the lamp
6. the Radio
7. the telephone
8. the computnre
9. the drier
10. the toster

B. Pick 1 thing that has a switch. Draw it.

Fifth-Grade Inventors Devise Intrusion Alarms and Code Makers *by Angel Gonzalez*

We have already visited Angel Gonzalez's science room at the Family Academy in Central Harlem. Angel regarded his students' investigations of retractable pens and valves as preparation for work with electrical controls. He began this unit by showing them how to make a complete circuit with a battery and a bulb. He then formulated another "situational challenge," which was modeled on the broken valve problem discussed earlier.

"Now that you have a circuit connected to a battery, can you come up with a device that will control the flow of 'electric juice' so that it does not get wasted? You are to invent an electric switch that can turn electricity on or off in a circuit, just like a faucet that controls the flow of water. The switch device must work in a consistent and a reliable fashion. It must not be the simple disconnecting and reconnecting of a wire, bulb, or battery to break the circuit."

With this problem, Angel introduced the idea that the flow of electricity is similar to the flow of water. In order to control it, a device is needed that is similar to a faucet or valve. Like Kathy, he wanted to expose them to a rich variety of devices, which would stimulate ideas about how they might proceed.

Various broken appliances and toys were made available to the students so they could dissect them and explore how the switches worked. The materials included keyboards, game controllers, flashlights, lamps, radios, tape players, computer "mouses" and more. I encouraged students to open up any switches they could, to see how the key parts function to regulate the flow of electricity. Some students made displays of the devices and shared how they thought the switches worked.

After some dissections and explorations of some switches, I asked the class to list as many switches as they could find in our classroom or in their homes. They were to list them and identify the type of action required to make each one work. After they had conducted discussions within their teams, I charted their collective findings, shown in Table 4-2.

We defined a switch as something that turns an electric device on or off. It is a control that regulates electric flow.

December 1

Brandon's team invented a switch by simply putting a metal paper fastener between a non-conducting part of the battery and the battery holder. As he pressed down on the fastener, the terminal of the battery separated from the metal on the battery holder and turned the bulb off. Quite clever!

December 2

Sophia asked, "What materials are inside the pull-string switch in the garden area? Wouldn't the inside of the switch burn with all that electricity and heat?"

I responded that those are important factors that must be considered in inventing switches. She could open the switches to see the materials they actually used to prevent overheating and fires.

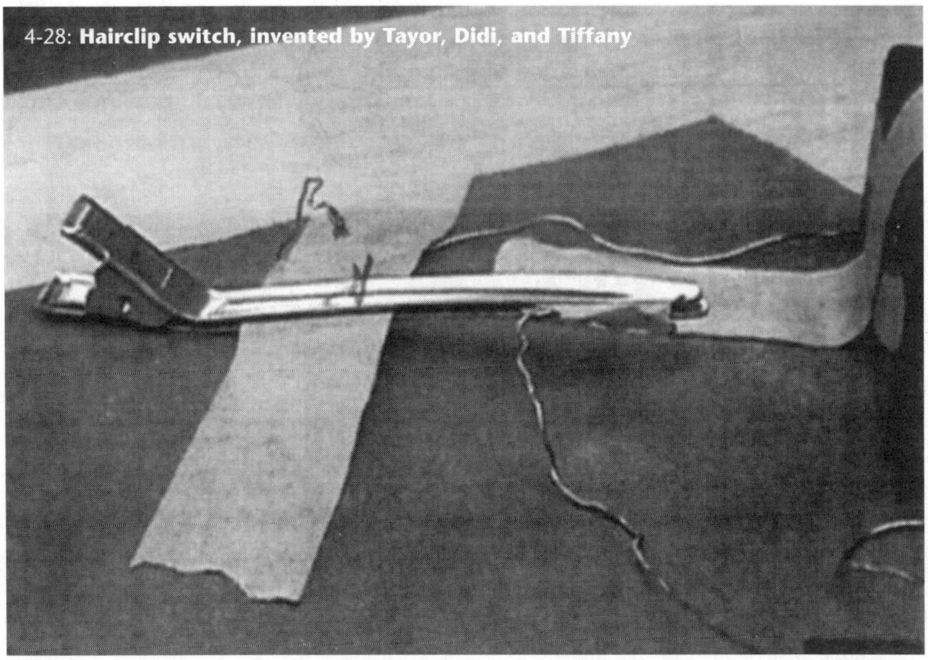

4-28: **Hairclip switch, invented by Tayor, Didi, and Tiffany**

December 7

The fifth-graders explored various switches. They also began to design circuits involving switches that they would have to invent themselves. I provided each group with a box of the supplies needed to make four complete battery-and-bulb circuits. Each box also contained the following: fasteners, paper clips, and index cards. These are materials they could use to invent simple switches, possibly with a few hints from me.

I let them know that they could bring in any materials they wanted, and that they could also give me a list of things they might need. Because they have already analyzed commonly found switches, some students feel confident enough to begin trying to invent their own switches. I suggested that they might think about designing flashlights or light circuits for a model home. They could also make exploded drawings of an appliance, highlighting the switch controls and explaining how they work.

December 8

The fifth-graders had only 40 minutes with me today, so I let them work on drawing up plans for their team projects. Table 4-3 shows a summary of the plans so far.

The class appears to be very excited about these projects. I will provide periodic research time during which they can explore books on electricity in my library corner. I strongly recommended the use of the CD-ROM and book, *The Way Things Work*. They can copy sections of the books that are relevant to their work. Those who need more time can come up during lunch to work on their projects.

Table 4-2

SWITCHES THAT ANGEL'S CLASS FOUND

Name of switch	Place found	Action required
Button	Computer	Pushing down
Flicker	Wall	Flick down or up
Game button	Play Station	Pressing
Turner	TV	Twisting right or left
Remote	Cable box	Pressing
Chain switch	Lamp	Pulling
Sound-sensitive	Gorilla doll	Loud-noise making
Slider	Flashlight	Sliding
Motion detector	Faucet	Hand motion

Table 4-3

PLAN SUMMARIES

Team	Ideas
Charmaine's	Blow dryer dissection and own invention
Shantale's	Lighting system for home
Taylor's	Burglar alarm
Santiago's	Fan
Hakim's	Exploded drawing of singing parrot doll
Dashaye's	Lighting system for castle
Sara's	Drawings of trackball; joystick invention
Christine's	"Squeeze Breeze" dissection
Sapphire's	Alarm system

December 15

Today, Eric and Ankaser invented a switch using aluminum foil, a paper clip, and a small card. Pressing down on the raised paper clip brought it into contact with the aluminum foil, completing the circuit and turning a motor on. They proudly shared their work.

December 21

Taylor's team invented a burglar alarm. They used a hair clip and a raised paper clip. When the paper clip was pressed, it touched the hair clip and completed the circuit, triggering a buzzer-alarm. The girls worked creatively with construction paper and hid the switch under a mat near the teacher's desk. When an unsuspecting person stepped on the mat, the buzzer would go off and draw attention to the intruder. Taylor, Didi, and Tiffany, the proud inventors, shared this device, shown in Figure 4-28, with the class.

Clearly, many of Angel's students became knowledgeable about electrical controls. They saw the analogy between electricity and fluid flow, looked inside a variety of appliances and switches to see how they worked, and found out how to interrupt the flow of electricity using a control device. They used this knowledge to build useful devices. At the same time, Angel felt that some of his students had been left behind. In retrospect, he felt that he should have preceded this activity with a science unit on electricity.

While some students were successful and had an easy time inventing their own switches, a good number were having difficulty and became frustrated in setting up complete circuits using light bulbs. It became apparent that they were not grounded enough in basic electricity concepts such as the flow of electricity, conductors/non-conductors, etc.

I therefore decided to spend the next month or two (January and February) introducing lessons from a science curriculum unit on electricity. The skills and concepts include simple circuits, conductors/nonconductors, how a bulb works, direction of flow of electricity, and series and parallel circuits. With a solid foundation, the students will be better able to take up the switch challenges.

During the next two months, Angel's fifth-grade science class went through the electricity unit that had been provided by the District. In March, Angel and his fifth-graders returned to the work with electrical controls. He came up with a project that combines the study of switches with another ***Stuff That Works!*** topic, *Signs, Symbols and Codes.*

He began with a problem that is very real for his class: the frequent interruptions from the main office or from elsewhere in the school. Could an electric code-sender be used to send messages between rooms, and thereby cut down on the number of interruptions? The students would not only have to design and test the devices for sending and receiving messages, but also develop the codes that would be used. Their inventions could actually be tested between rooms, and put in place if the Principal agreed.

The Challenge

Each student will design an encoder with a signaling switch, and show and explain the design in his/her journal. Each student will share ideas with the rest of the team, and the team will adopt one plan to implement. Then the teams will gather the necessary materials, and proceed to build and test the design.

March 1

I gave the fifth-graders the Encoder Challenge and most were excited about it. They drew their designs and wrote up their plans in their journals. They have to share their designs and adopt one as a team. Immediately, some began to develop codes to use with their systems. I suggested that they could use either bulbs or buzzers to convey the signals. I also said that they could divide up the work within the teams. Some could work on the encoder, while others worked on the codes and their meanings.

Lauren shared her team's idea using the overhead projector:

"We'll get that wire that goes from the office. Our Principal can put this button on her desk and it goes to every classroom in the Family Academy.

- If it lights two times, we got school sing;
- If it lights three times, fire drill; and
- If it lights four times, the principal wants to have a meeting."

Samantha presented this design, shown in Figure 4-29, from her team:

"There will be a buzzer with a button and a bulb. It will be connected to a battery, a big one. Then there will be three wires, one to the battery, one to the bulb, and one to the buzzer. The wires will be connected to all the rooms in the school. If the bulb lights up, that's for emergencies like fire drills. You press down on the buzzer and it sends to the office. One buzz is any announcement. Two beeps is 'Come pick up this child.' Three beeps is emergency, and if it's a fire drill, the office will have a light bulb. It will send to our room and then we will know."

During lunch hour, Brittney came upstairs with her teammates, Charmaine and Antemia, to finish and test out their buzzer encoder. They added longer wires and succeeded in sending their coded messages. The next day they shared their work with the entire class.

This account confirms the soundness of Angel's decision to develop the students' background on circuits before proceeding further with electrical controls. It is clear from this account that the students felt confident about circuits, and were able to use their knowledge to solve practical problems.

4-29: **Samantha's sketch of the electric code-sender system**

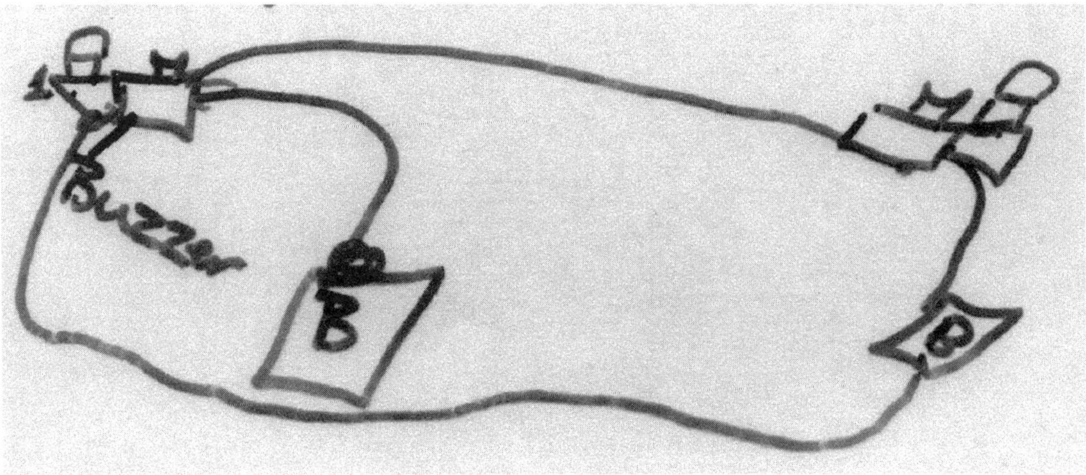

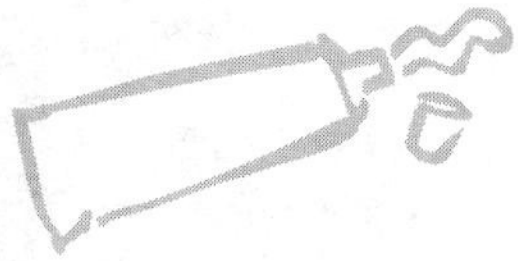

Chapter 5

RESOURCES

Help for Teachers

Making Connections with Literature

Using literature as a supplement and enhancement for instruction is good teaching practice because:

- Children learn from everything they experience.
- Children learn more effectively when instruction is associated with positive emotions, such as those evoked by a good book.
- Literacy is key to children's success as learners.
- There are many different learning styles.

We encourage you to incorporate books of all kinds into your work with *Mechanisms and Other Systems.* We've included an annotated list of quality books of all kinds on the following pages—storybooks in which the use of mechanisms and other systems are demonstrated, as well as nonfiction books on how things work, how to make things, how to conduct experiments, and how some people have had bright ideas that led to inventions that help us in our everyday lives.

But don't stop with these. You know your students and how they learn better than anyone else. When you see a book that might further your instructional goals, interest or challenge a particular student, or evoke feelings that make learning more fun, add it to the books that are available to your students.

Action Robots: A Pop-up Book Showing How They Work, by Tim Reeve. Dial Books: New York, 1995. (Recommended grades: 2-6)

Fact-filled pop-up book describes the many kinds of robots that are being used today in science, industry, and medicine. Each page has a three-dimensional figure with tabs that can be pushed, pulled, or moved to simulate the actions of these remarkable machines. The book itself is a mechanism!

Alexander and the Wind-up Mouse, by Leo Lionni. Dragonfly Books/Alfred A. Knopf: New York, 1974. (Recommended grades: PreK-3)

A fantasy about the friendship between a real mouse and a mechanical mouse. Enjoyable language arts extension for hands-on mechanism processing and inquiry-based questioning.

Alexander Graham Bell: Making Connections, by Naomi Pasachoff. Oxford University Press: New York, 1998. (Recommended grades 4-6)

This compelling biography of a true scientific visionary charts the course of Alexander Graham Bell's remarkable life, showing how his early studies of speech and sound and his experience as an instructor of the deaf led to his invention of the telephone in 1876.

The Berenstain Bears Go In and Out, by Stan Berenstain and Jan Berenstain. Random House: New York, 2000. (Recommended grades: K-3)

Going out to eat fast food isn't fast at all when the Berenstain Bears use the revolving door. This simple "Early Steps into Reading" book integrates mechanisms and fosters reading for beginners.

The Berenstain Bears' Science Fair, by Stan Berenstain and Jan Berenstain. Random House: New York, 1984. (Recommended grades: K-4)

Papa Bear teaches Small Bear and Sister about machines and how they work and helps them prepare projects for a science fair. Information is presented in simple terms and lively pictures. An enjoyable, concrete extension to mechanisms for younger students.

The Best of Rube Goldberg, by Charles Keller. Prentice Hall: Englewood Cliffs, NJ, 1979. (Recommended grades: 4 and up, Teacher Resource)

Contains a collection of more than 90 humorous inventions, which have appeared in Rube Goldberg's cartoons during his career. For 55 years, this cartoonist drew machines and contraptions of marvelous complexity and ingenuity to perform such basic tasks as scratching your back or hitting a mosquito. His mechanical inventions have fascinated, amused, and challenged people.

Batteries, Bulbs, and Wires, by David Glover. Kingfisher: New York, 1993. (Recommended grades: 2-6)
Activities and projects introduce how electricity and magnets work at home and in the everyday world. Filled with bright, clear photographs with at least one activity to illustrate each concept.

Cars And How They Go, by Joanna Cole. Thomas Y. Crowell: New York, 1983. (Recommended grades: 2-6)
Turn the key, shift the gears, step on the gas pedal, and drive away. It all seems simple until you look under the hood. Dozens of working mechanisms, push and lift, slide and turn, to make the power that makes the wheels go round. Reader-friendly visuals provide for easy comprehension of a car's mechanisms.

Construction for Children, by Barbara Eichelberger and Connie Larson. Dale Seymour, 1996.
A collection of activities focusing on problem solving, constructing, and evaluating working models based on themes from literature.

Cutaway Trains: Look Inside Machines to See How They Work, by Jon Richards. Millbrook Press: Brookfield, CT, 1998. (Recommended grades: K-3)
Presents different kinds of trains, including steam trains, diesel locomotives, and electric monorails, and discusses how they perform various jobs. Detailed full-color illustrations peel off the outer layers to reveal the complex technology and mechanisms within some of the fastest and most powerful trains in the world.

Danny Dunn and the Swamp Monster, by Jay Williams. McGraw-Hill: New York, 1971. (Recommended grades: 4-6)
Danny and his friends discover a superconductor, which they use in an adventure in Africa involving an electric fish. Excellent explanations of electrical circuits, current, resistance, and magnetic fields are provided in the story.

Dear Mr. Henshaw, by Beverly Cleary. William Morrow & Co.: New York, 1994. (Recommended grades: 3-6)
In his letters to his favorite author, ten-year-old Leigh reveals his problems in coping with his parent's divorce, being the new boy in school, and finding his own place in the world. A portion of the book describes how the boy sets out to catch a thief by rigging a battery-powered burglar alarm to his lunch box, and how this invention gains him respect. An outstanding literary link for developing ELA skills while integrating circuits with literature.

Doctor De Soto, by William Steig. Farrar, Straus & Giroux: New York, 1990. (Recommended grades: K – 3)
This diminutive resourceful mouse dentist copes with the toothaches of various animals. He utilizes various mechanical devices in his practice to accommodate his patients. Excellent literature link for introducing mechanisms to young students, while integrating inquiry-based questioning and ELA skill development.

Electric Gadgets and Gismos, by Alan Bartholomew, Kids Can Press: Toronto, 1998. (Recommended grades: 4-6)
Battery-powered easily built gadgets that go. Clear step-by-step instructions and detailed color illustrations. Creative fun with hands-on nechanisms.

Electricity, by Peter Riley. Franklin Watts: New York, 1999. (Recommended grades: 3-6)
Explains simple circuits, generating electricity, motors and their uses in everyday life. Presents hands-on experiments.

Garrett Morgan, Inventor, by Garnet N. Jackson. Modern Curriculum: Cleveland, 1992. (Recommended grades: K-4)
A brief biography of the African American inventor that describes, in rhyming text, how his lifelong interest in machines led to the invention of the traffic signal. An excellent interdisciplinary literature link for mechanisms integrating technology, social studies and African American history.

Get Ready for Robots! by Patricia Lauber. HarperCollins: New York, 1987. (Recommended grades: 4-6)
An introduction to robots, describing clearly and accurately the tasks they can perform at home and in industry, and the things they might do in the future. Explains the limitations of what robots can be programmed to "see" and "do," and why some robots still need adult control.

How Do You Lift a Lion? by Robert E. Wells. Albert Whitman: New York, 1996. (Recommended grades: 2-6)
Basic introduction to levers, wheels, and pulleys. It aims to introduce some principles of physics to young readers. The concepts of gravity, leverage, friction, and pulleys are illustrated accompanied by some physical problems involving animals. A very clever book for introducing young students to mechanisms.

Levers, by Michael Dahl. Capstone Press: Mankato, MN, 1996. (Recommended grades: K-3)
Describes the many kinds, uses, and benefits of levers. Lots of visuals with text. A helpful word-to-know list for students is in the back of book.

Liz On The Move (The Magic School Bus), by Joanna Cole and Bruce Degen. Scholastic, Inc.: New York, 1999. (Recommended grades: K-6)
Wanda has a new tree house, and she is ready to move in with all her favorite things. Liz helps her friend get the job done in no time with a handy clothesline pulley, ramps, and wheels. An enjoyable look at using simple machines to make a task easier.

Machines, by Janice Van Cleave. John Wiley & Sons, Inc.: New York, 1993. (Recommended grades: 3-6)
A collection of science projects and experiments that explore simple machines. Primary grade students can use this link for projects with adult supervision. Upper grade students can be involved independently. A good teacher resource for classroom projects.

Machines and How They Work, by David Burnie. DK Publishing: New York, 1991. (Recommended grades: 3-6+)
Begins by defining machines and then discusses early machines, both simple and complex. Pictures like "Time and Motion" and "Bicycles" show the development of the modern machine. Mechanisms come alive as sundials, wristwatches, water wheels, modern hydroelectric generators, sewing machines, and cars are illustrated.

Machines And How They Work, by Harvey Weiss. Thomas Y. Crowell: New York, 1983. (Recommended grades: 4-6)

An introduction to six simple machines, and their use in more complex machines such as derricks, bulldozers, and metal lathes. Historical background material is provided along with modern examples. Reinforces understanding of basic physics principles!

Machines as Tall as Giants, by Paul Strickland. Random House: New York, 1989. (Recommended grades: K-3)

A variety of very large machines, including tower cranes, a dockyard crane, oil production platform, space shuttle, and a concrete silo are depicted on two pages sideways to emphasize heights. Easy to read text accompanies each picture.

The Magic School Bus and the Electric Field Trip, by Joanna Cole and Bruce Degen, Scholastic Inc.: New York, 1998. (Recommended grades: 1- 6)

Ms. Frizzle takes her class on a field trip through the town's electrical wires so they can learn how electricity is generated and how it is used. Many facets of electricity from circuits and switches to computers are covered.

Marbles, Roller Skates, Doorknobs: Simple Machines That Are Really Wheels, by Christopher Lampton. Millbrook Press: Brookfield, CT., 1994. (Recommended grades: 2 – 6)

Discusses the physical principles involved in the functioning of wheels. Explains how simple machines work that are actually wheels. Included examples are machines that are familiar to children such as a pizza cutter, a bicycle, and a water faucet. Well illustrated and easily comprehended.

Mike Mulligan and his Steam Shovel, by Virginia Lee Burton. Houghton Mifflin Company: Boston, 1977. (Recommended grades: K-3)

A fun book both in its text and gray crayon drawings! Mike Mulligan remains faithful to his steam shovel Mary Anne, against the threat of the new gas and diesel-engine contraptions. He digs his way to a surprising and happy ending.

National Geographic: How Things Work, by John Langone. National Geographic Society: Washington, D.C., 1999. (Recommended grades: 4-6+; also a teacher resource)

A fascinating and clearly written book that helps the reader understand the mechanisms and principles behind machines, processes, and systems. Stunning photographs and diagrams illustrate this book.

The New Way Things Work, by David Macaulay. Houghton Mifflin Company: New York , 1998. (Recommended grades: 5+/ Teacher resource)

From levers to lasers, windmills to web sites, a visual guide to the world of machines. This revised edition includes a new section devoted to digital machinery.

On Wheels, by Huck Scarry. Philomel Books: New York, 1980.

Join caveman Peter Pebble on a journey through time and space to observe the development of the wheel from its probable discovery in the Stone Age to the early 1900's and gasoline powered "horseless carriages." Also included are such wheel-related processes as road and bridge building, blacksmithing, and harness making. Integrates American history, world history, and social studies.

Paper Engineering for Pop-Up Books and Cards, by Mark Hiner. Parkwest Publishers: 1986.

Illustrates the ten basic mechanisms, which are used in making pop-up books. This book is a gem for teachers interested in acquainting children with the engineering principles of pop-ups both as an appreciation of books they enjoy and as an introduction to making their own. A great resource for hands-on mechanisms!

The Pop-Up Book of Big Trucks, by Peter Seymour. Little, Brown & Co.: Boston, 1989. (Recommended grades: 1-3)

Using a variety of moving parts, pull tabs, lift ups, die cuts, slides and hinges, this unique pop-up book involves readers in the workings of all sorts of big trucks. You can raise and extend the ladder, lift the tailgate, dump the sand, and turn the barrel on the cement mixer.

Projects With Machines, by John Williams. Gareth Stevens Children's Books: Milwaukee, 1990. (Recommended grades: 2-6)

Provides instruction for making and using a variety of machines, including levers, diggers, cranes, and pulleys.

Pulleys, by Michael Dahl. Capstone Press: Mankato, MN, 1999. (Recommended grades: K-3)

Easy-to-read science book that describes the many different kinds, uses, and benefits of pulleys. Full color photographs are tied closely to text. Important words to know are listed at the end of the book.

Richard Scarry's Pop Up Wheels, by Richard Scarry. Simon & Schuster: New York, 1997. (Recommended grades: PreK-2)

This book is bursting with vehicles of all shapes and sizes. Hands-on connection for mechanisms with this pop-up!

Roller Coaster Science: 50 Wet, Wacky, Wild, Dizzy Experiments About Things Kids Like Best, by Jim Wiese. John Wiley & Son: New York, 1994. (Recommended grades: 4 and up)

This entertaining and educational book describes the science behind such amusements as roller coasters, swings, bumper cars, curve balls, and more. Contains numerous activities and experiments to help reinforce scientific principles and covers a variety of disciplines including math, chemistry and physics. Great resource link for hands-on work with mechanisms.

Seesaws, Nutcrackers, Brooms: Simple Machines That Are Really Levers, by Christopher Lampton. Millbrook Press: Brookfield, CT, 1994. (Recommended grades: 3-6)

Explains the principles involved in the functioning of levers, and how these simple machines are used in daily life. Graphics provide visual reinforcement.

Simple Machines, by Michael S. Dahl. Capstone Press: Mankato, MN, 1996. (Recommended grades: K-3)

Easy-to-read science book introduces young readers to simple machines, with full-color visuals tied closely to the text. The book provides basic information about levers, inclined planes, pulleys, and wheels and axles.

Simple Machines, by Deborah Hodge and Ray Boudreau. Kids Can Press Ltd.: Toronto, 1998. (Recommended grades: K-4)

Thirteen hands-on experiments are clearly explained using full-color photographs and step-by-step directions. Extensions of each activity are included in an appendix.

Springs, by David M. Glover. Heinemann: Portsmouth, NH, 1997. (Recommended grades: 2-6)

How are a jack-in-the-box, a door lock, and an umbrella alike? How can springs help you get somewhere on time? These questions and others are answered, and the book explains how different kinds of simple machines are all around us, making our work and play easier. Visuals provide easy comprehension of mechanisms and their purposes.

Switch On, Switch Off, by Melvin Berger. HarperCollins: New York, 1990. (Recommended grades: K-3)

The magic of a light switch is made accessible to everyone with this easy-to-read introduction to electricity, from the generator to the individual light bulb. The book includes an experiment in which electricity can be generated using just a magnet, a compass, and a piece of wire.

Tom Edison's Bright Idea, by Jack Keller. Steck-Vaugh: Austin, TX, 1992. (Recommended grades: 2-6)

A biography for beginning readers, focusing on the inventor's research with electricity and circuits and his invention of the incandescent light bulb.

Until I Met Dudley: How Everyday Things Really Work, by Roger McGough. Walker & Company: 1997. (Recommended grades: 1-6)

A young girl used to have fantastic ideas about how things work, but Dudley, a bespectacled dog, tells her how it really is. He explains the workings of mechanical objects such as vacuum cleaners, refrigerators, dishwashers, toasters and garbage trucks. Each explanation is preceded by a fantasized description and illustration of how things might work in the world of imagination. Great for developing critical thinking skills.

What Was It Like Before Electricity, by Paul Bennett. Steck-Vaughn: Austin,TX, 1995. (Recommended grades: 2-6)

On a visit to great-grandmother, a boy and his friends learn about daily life and mechanical power before the availability of electricity. This book is easy to read independently, and provides an informative literary link for integrating science, technology, and language arts.

The Wheeling and Whirling Around Book, by Judy Hindley. Candlewick Press: Cambridge, MA, 1994. (Recommended grades: K-3)

This book explores the virtues of circles and disks and spirals and spheres, of rolling and spinning and turning and churning. Some of the things it describes are merry-go-rounds, gears, axles and conveyor belts. It combines exciting stories, poems, and artwork with elements of the best non-fiction.

Wheels and Axles, by Michael S. Dahl. Capstone Press: Mankato, MN, 1996.
(Recommended grades: K-3)

An easy-to-read science book describing the many different kinds, uses, and benefits of wheels and axles. Full color photos are closely tied to text.

Wheels and Cranks, by David Glover. Rigby Education: Crystal Lake, IL, 1997.
(Recommended grades 2-5)

Introduces the principles of wheels and cranks as simple machines, using examples from everyday life. Great visuals.

Wheels At Work, by Bernie Zubrowski. William Morrow & Company: New York, 1986.
(Recommended grades: 5 and up)

A "Boston Children's Museum Activity Book" lets you find out how useful and versatile the ordinary wheel is, with over 50 hands-on experiments. Helpful resource for classroom teachers looking for fresh ideas in extending mechanisms activities.

Why Didn't I Think of That?, by Webb Garrison. Prentice-Hall: Englewood Cliffs, NJ, 1977.
(Recommended grades: 5 and up)

From alarm clocks to zippers, ballpoint pens, cash registers, elevators, Lionel trains, safety pins, sewing machines, windshield wipers, Yale locks, and more. Every one of these inventions came to its inventor as a brainstorm. A good resource for extended activities.

Assessment

Nearly everyone agrees about the importance of assessment, but what exactly is it, and why is it so significant in education? In a very broad sense, education is like a very large design problem and assessment is the method of evaluating the design. However, education has many objectives, not just one, so assessment also includes a complex process of deciding what to assess and how. Another major complication is that many different kinds of people have a stake in the outcome of the educational process. Parents want to know how much their children are learning and how they can best help them. Politicians worry about the backlash from voters if the educational system appears to be "failing," however that term is defined. Administrators fear that they will be held accountable for low test scores in their schools.

Teachers, who have the most sustained and direct involvement of any adults in the educational process, are constantly looking for ways of knowing how well and how much their students are learning. This data can come from both formal and informal assessment methods, and may be either qualitative or quantitative. At the same time, teachers are often held accountable to conflicting requirements that are difficult or impossible to meet. For example, the goal of providing a supportive and welcoming learning environment may be in conflict with the regimentation imposed by administrative requirements. Another common concern of teachers is that high-stakes testing will require them to "teach to the test" rather than to support student learning.

Regardless of demands from outside the classroom, a teacher's primary responsibility is to engage students in exploring and understanding the subject matter. Assessment includes any method of finding out how much of this exploring and understanding actually happens. Information gained through assessment is the only factual basis for knowing what students are learning, how to motivate learning more effectively, how and whether to redesign the curriculum, how to tailor it to the needs of individual students, and how and when to involve parents in the process. Assessment is far too extensive and important to be narrowly defined by standardized test results or to be determined by people outside the classroom.

Here are some basic conclusions that follow from this view of assessment:

- Assessment should be based on clear educational goals.
- Many different kinds of information should be collected as part of assessment. Some of the most important assessment data is totally unexpected.
- Assessment should not be divorced from curriculum; every learning activity should also provide information for assessment.
- Whenever possible, students should become involved in assessing their own learning—for example, by evaluating their own designs or predictions.
- Assessment should examine not only what students have learned, but also the opportunities provided by the curriculum and the learning environment.

We will illustrate each of these points using examples from the teacher stories in Chapter 4 of this guide.

Educational Goals

In order to assess the learning outcomes of an activity, it is necessary to know what the educational goals were. However, the purpose of a curriculum unit may not be so clear-cut. Any worthwhile educational activity probably has more than one goal. Also, a teacher's goals may (and often do) change as the activity progresses, or there may be unintended outcomes that are far more significant than the original goals.

In thinking about assessment, it is important to regard educational goals in a broad, flexible way. For example, when Mary Flores began her unit on mechanisms, her primary goals were for her students to understand the concept of a mechanism, see some of the purposes of mechanisms, and be able to identify simple machines within more complex mechanisms. As the unit unfolded, however, a new issue developed that was at least as significant as any of the original intentions. The culminating activity was to be the design of "Rube Goldberg" devices using the simple machines the students had studied. Quite by accident, the custodian discarded some of the students' initial designs. Although Mary was crestfallen, the students saw this mishap as an opportunity to redo and improve upon their original designs. Thus the focus of the unit became the design process itself, and the possibility of revisiting an initial design and making it better.

Rigid adherence to an initial set of goals assumes that the educational process is entirely predictable, which is not the case. Every teacher has both short- and long-term goals for her students, and it is difficult to know in advance when something will happen to advance the long-term goals unexpectedly. As one teacher put it during a discussion on assessment, "You can talk about goals all you want, but what I really care about is that they feel good about themselves and about what they are able to accomplish."

Information from a Variety of Sources

If educational goals are complex and multifaceted, so are the means of assessing to what extent these goals are met. The narrowest view of assessment, most popular in political circles, confines it to standardized tests. A somewhat broader view expands assessment to include all kinds of paper-and-pencil instruments designed specifically for assessment, such as worksheets, homework assignments, tests, and quizzes.

Our view of assessment is broader still. In the course of an activity, nearly anything students do generates information that is valuable for assessment. To give just one example, when students talk about their work, they often provide useful data about the learning process. Mary's student Priscilla, for instance, provided this little gem of information about how she was grappling with the design process: "It's making me angry, because I tried hard tomake my mechanisms work, but now I have to make another one."

Part of the attraction of teaching is that much of what happens in the classroom is unpredictable, and some of the surprises are pleasant and even thrilling. Consequently, it is impossible to decide in advance what all of the methods of assessment will be. Often, serendipity provides ways of assessing students' learning that nobody could have anticipated. A dramatic example, already cited, occurred when the custodian threw out Mary's students' work, forcing them to redesign it and reflect on the design process. More mundane examples happen frequently. Unexpectedly, a child may make a comment or ask a question that indicates reflection about the concepts and connections made among them. For example, one of Angels' students asked, "What materials are inside the pull-string switch in the garden area? Wouldn't the inside of the switch burn with all that electricity and heat?"

Curriculum as a Major Source of Assessment Data

In order to maximize the amount of information available, the curriculum itself must be seen as a rich source of assessment data. Each of the teachers whose stories appears in Chapter 4 of this Guide used a brainstorming activity as part of a unit. Annette Purcell asked her first-graders for examples of tools and what they do; Mary Flores and Shirley Nieves asked their special education students to cite mechanisms from their own experience; Kathy Aguiar elicited examples of controls from her third-grade special education students; and Angel Gonzalez requested instances of switches from his fifth-grade science students.

Each of these brainstorming sessions was both a worthwhile educational activity and also a valuable opportunity to assess the students' prior knowledge about the topic. Similarly, when Shirley's students sorted mechanisms by categories, they were both learning about mechanisms and providing their teacher with useful assessment data. The same could be said of virtually any activity associated with a curriculum unit. Model making, design activities, presentations to the class, journals, and discussion within a work group are all potential sources of assessment information.

Students Assess Their Own Learning

Should the audience for assessment data include students themselves? Obviously, students need to know how well they are doing, so they can gauge their own efforts and develop realistic goals for their own learning. However, traditional assessment is usually presented to students in an adversarial manner, in the form of test grades and report cards that frequently undermine rather than enhance their motivation for learning. In traditional forms of assessment, students are always evaluated by adults rather than by themselves, and the outcomes of assessment often have high stakes. Both of these factors contribute to the view of assessment as an antagonistic process. How can students gain access to candid data about their own learning without interpreting it as somehow the product of bad intentions?

A way out of this dilemma is suggested by some of the activities in Chapter 4. Technological analysis and design activities often provide occasions for self-assessment, where students evaluate their own work against an objective standard rather than one arbitrarily set by adults. Angel's students, for example, discussed the role of switches in circuits by making the analogy with the role of valves in controlling the flow of water. Then they took apart and examined some switches to see how they interrupt and restore the flow of electricity when turned off and on, respectively. As a culminating activity, they had to design their own switches and incorporate them into systems such as intrusion alarms. This design project was also a form of self-assessment. As the students tested their own designs to see if they worked, they were also evaluating how much they had learned about switches and whether they had been able to apply this knowledge to a practical design problem.

Assessing the Learning Environment

Like anybody else who designs or plans anything, most teachers engage in informal assessment of their work on an ongoing basis. They ask themselves, "Is it working?" This question is really one of self-assessment: "What is the quality of the learning opportunities I have provided for my students?"

Some of this self-assessment by teachers is based on student learning outcomes of the many kinds described above. At the same time, teachers also assess learning opportunities on the basis of their own perceptions and experiences. Several examples of these self-assessments appear in the teachers' stories in Chapter 4. Annette, for instance, states, "Several children found the movement of the scissors confusing. I would change the presentation by suggesting that they keep one part of the scissors fixed. The stationary part would make a better frame of reference for seeing the movement of the other part."

Mary actually revised her activity in midstream based on a self-assessment: "I instructed the morning students to begin grouping the mechanisms. Things did not go smoothly. They did not understand the term grouping. I had to think quickly. How could I get them to understand the concept?" By the end of the day, she had invented an activity on grouping in which students categorized themselves based on their similarities and differences.

Angel went even further. Finding that some students "were having difficulty and became frustrated in setting up complete circuits," he decided to interrupt the work on controls and switches. His students spent the next two months doing a "batteries and bulbs" unit from a science curriculum before returning to electrical controls later in the year.

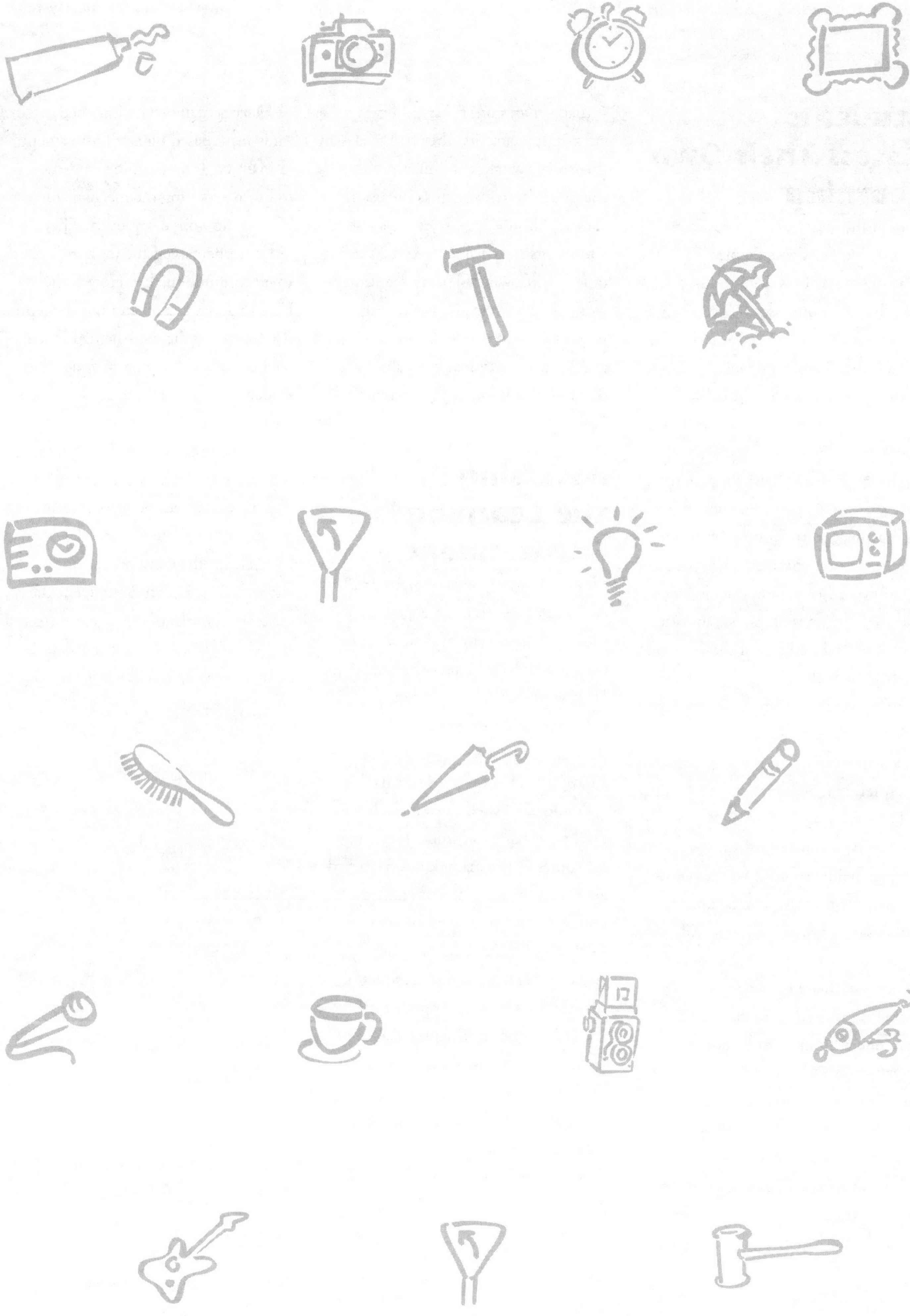

Chapter 6

ABOUT STANDARDS

Overview

In Chapter 3, "Activities," we have listed standards references for each activity. This type of listing is now found in most curriculum materials, in order to demonstrate that the activities "meet standards." In a way, these standards references miss the point, because the national standards are not meant to be read in this way. Meeting standards is not about checking off items from a list. Each of the major standards documents is a coherent, comprehensive call for systematic change in education.

This chapter shows how ***Stuff That Works!*** in general and *Mechanisms and Other Systems* in particular are consistent with national standards at a very fundamental level. We will look in detail at the following documents:

- *Standards for Technological Literacy: Content for the Study of Technology* (International Technology Education Association, 2000);
- *Benchmarks for Science Literacy* (American Association for the Advancement of Science, 1993);
- *National Science Education Standards* (National Research Council, 1996);
- *Standards and Principles for School Mathematics* (National Council of Teachers of Mathematics, 2000);
- *Standards for the English Language Arts* (National Council of Teachers of English & International Reading Association, 1996); and

Most of these standards are now widely accepted as the basis for state and local curriculum frameworks. The first document on the list is included because it is the only national standard focused primarily on technology. The *New Standards Performance Standards* (National Center on Education and the Economy, 1997) is not included because it is based almost entirely on the *Benchmarks, National Science Education Standards,* the original NCTM *Math Standards* (1989) and the *Standards for the English Language Arts.*

Although they deal with very different disciplines, these major national standards documents have many remarkable similarities:

- They are aimed at *all* students, not only those who are college-bound.
- Using terms like "literacy" and "informed citizen," they argue that education should prepare students to understand current issues and participate in contemporary society.
- They recommend that school knowledge be developed for its use in solving real problems rather than as material "needed" for passing a test. They strongly endorse curriculum projects that arise from students' own ideas, experiences, and interests.
- They focus on the "big ideas" of their disciplines as opposed to memorization of isolated facts or training in narrowly defined skills. In other words, fewer concepts should be dealt with in greater depth. As the *National Science Education Standards* express it, "Coverage of great amounts of trivial, unconnected information must be eliminated from the curriculum." (NRC, 1996, p. 213)

- The standards reject standardized tests as the sole or even the major form of assessment. Traditional exams measure only what is easy to measure rather than what is most important. "While many teachers wish to gauge their students' learning using performance-based assessment, they find that preparing students for machine-scored tests – which often focus on isolated skills rather than contextualized learning – diverts valuable classroom time away from actual performance." (NCTE/IRA, 1996, p. 7) The standards promote authentic assessment measures, which require students to apply knowledge and reasoning "to situations similar to those they will encounter outside the classroom." (NRC, 1996, p. 78) Furthermore, assessment should become "a routine part of the ongoing classroom activity rather than an interruption" and it should consist of "a convergence of evidence from different sources." (NCTM, 2000, p. 23)

- They highlight the roles of quantitative thinking, as well as oral and written communication, in learning any subject, and they emphasize the interdisciplinary character of knowledge.

- They view learning as an active process requiring student engagement with the material and subject to frequent reflection and evaluation by both teacher and learner.

- They urge teachers to "display and demand respect for the diverse ideas, skills and experiences of all students," and to "enable students to have a significant voice in decisions about the content and context of their work." (NRC, 1996, p. 46)

The ***Stuff That Works!*** materials are based on these ideas and provide extensive guidance on how to implement them in the classroom. We begin our study of technology with students' own ideas and experiences, address problems that are of importance to them, develop "big ideas" through active engagement in analysis and design, and draw connections among the disciplines. While the standards are clear about the principles, they do not provide many practical classroom examples. ***Stuff That Works!*** fills this gap.

Where the Standards Came From

Historically speaking, the current tilt towards national curriculum standards is a dramatic departure from a long tradition of local control of education. How did national standards manage to become the order of the day? In the late 1970's, the country was in a serious recession, driven partly by economic competition from Europe and Japan. In 1983, the National Commission on Educational Excellence (NCEE) published an influential report, *A Nation at Risk,* which painted a depressing picture of low achievement among the country's students. The report warned of further economic consequences should these problems continue to be ignored, and advocated national curriculum standards for all students. Adding to these arguments were pressures from textbook publishers, who felt that national standards would make state and local adoption processes more predictable.

Around the same time, several of the major professional organizations decided to provide leadership in setting standards. The pioneering organizations were the National Council of Teachers of Mathematics (NCTM) and the American Association for the Advancement of Science (AAAS), whose efforts culminated in the publication of major documents in 1989. In the same year, the National Governors' Association and the first

Bush Administration both endorsed the concept of establishing national educational goals. The NCTM was deeply concerned about the issues raised by *A Nation at Risk* and was convinced that professional educators needed to take the initiative in setting a new educational agenda. Otherwise, the reform of curriculum would rest in the hands of textbook and test publishers, legislatures, and local districts.

Both the NCTM and the AAAS standards projects began with similar basic positions about pedagogy. Influenced by research about what children actually know, they recognized the disturbing fact that "learning is not necessarily an outcome of teaching." (AAAS, 1989, p. 145) In contrast with traditional approaches to education, which emphasize memorization and drill, the new national standards promote strategies for active learning. A related theme of the early standards efforts was that the schools should teach fewer topics in order that "students end up with richer insights and deeper understandings than they could hope to gain from a superficial exposure to more topics..." (p. 20)

Meeting standards requires a major investment of time and resources. Some of the necessary ingredients include new curriculum ideas and materials, professional development opportunities, new assessment methods, and smaller class sizes. The *National Science Education Standards* are the most explicit in identifying the conditions necessary—at the classroom, school, district, and larger political levels—for standards to be meaningful. The authors state, "Students could not achieve standards in most of today's schools." (NRC, 1996, p. 13) More money might not even be the hardest part. Standards-based reforms also require understanding and commitment from everyone connected with the educational system, starting at the top.

The history of standards may contain clues about their future. Standards imply neither textbook-based instruction nor standardized tests. Standards arose because traditional text- and test-based education had failed to result in the learning of basic concepts by the vast majority of students. Ironically, there are many textbook and test purveyors who market their products under the slogan "standards-based." Standards could easily become discredited if those who claim their imprimatur ignore their basic message.

What the Standards Actually Mean

Standards are commonly read as lists of goals to be achieved through an activity or a curriculum. This approach is reflected in the lists of standards references and cross-references that appear in most curriculum materials, as evidence that an activity or curriculum "meets standards." In the "Activities" chapter of *Mechanisms and Other Systems,* for example, we have listed the following reference under the activity "What Does a Tool Do?":

> "By the end of second grade, students should know that tools are used to do things better or more easily and to do things that could not otherwise be done at all." (AAAS, 1993, p. 45)

Presenting lists of outcomes in this fashion reflects a narrow reading of standards, which can be very misleading. These lists suggest that "meeting standards" is simply a matter of getting students to repeat something like the statements found in the standards documents, such as the one quoted above.

In fact, the standards are much richer and more complex than these lists imply. Many of the standards do not even specify the knowledge that students should acquire, but deal rather with ways of using that knowledge. Here is another example from *Benchmarks for Science Literacy:*

"By the end of fifth grade, students should be able to write instructions that students can follow in carrying out a procedure." (p. 296)

This standard talks about something students should be *able to do,* rather than what they should *know.* The newly released NCTM document, *Principles and Standards for School Mathematics* (2000), unlike the earlier one (NCTM, 1989), explicitly separates "Content Standards" from "Process Standards." The Content Standards outline what students should learn, while the Process Standards cite ways of acquiring and expressing the content knowledge. The Process Standards include problem solving, communication, and representation. The *Benchmarks* example just cited above is another example of a process standard. Similarly, in the English Language Arts (ELA) document (NCTE/IRA, 1996), all twelve standards use verbs to express what students should *do,* as opposed to what they should *know.* Any reading of standards that focuses only on content knowledge is missing a central theme of all of the major documents.

There is also material in the standards that qualifies neither as content nor as process. Here is an example from the *Benchmarks* chapter called "Values and Attitudes":

"By the end of fifth grade, students should raise questions about the world around them and be willing to seek answers to some of them by making careful observations and trying things out." (p. 285)

This standard asks for more than a specific piece of knowledge, ability, or skill. It calls for a way of looking at the world, a general conceptual framework, that transcends the boundaries of disciplines. Similarly, the "Connections" standard in the new NCTM document underscores the need for students to ...

"...recognize and apply mathematics in contexts outside of mathematics." (NCTM, 2000, p. 65)

These are examples of broad curriculum principles that cut across the more specific content and process standards. These standards are not met by implementing a particular activity or by teaching one or another lesson. They require an imaginative search for opportunities based on a reshaping of goals for the entire curriculum.

In general, the standards documents are at least as much about general principles as about particular skills and knowledge bases. The *Standards for Technological Literacy,* the *Benchmarks,* and the *National Science Education Standards* each identifies some big ideas that recur frequently and provide explanatory power throughout science and technology. "Systems" and "modeling" are concepts that appear in all three documents. The presence of such unifying ideas suggests that the individual standards references should not be isolated from one another. They should rather be seen as parts of a whole, reflecting a few basic common themes.

What Use Are Standards?

Increasingly, teachers are being held accountable for "teaching to standards." These demands are added to such other burdens as paperwork, test schedules, classroom interruptions, inadequate space and budgets, arbitrary changes in class roster, etc. In the view of many teachers, children and their education are routinely placed dead last on the priority list of school officials. Understandably, teachers may resent or even resist calls to "meet standards" or demonstrate that their curricula are "standards-bearing." It is not surprising that many teachers cynically view the standards movement as "another new thing that will eventually blow over."

The push to "meet standards" is often based on a misreading of standards as lists of topics to be "covered" or new tests to be administered. It is not hard to imagine where this misinterpretation might lead. If the proof of standards is that students will pass tests, and students fail them nevertheless, then the standards themselves may eventually be discarded. Paradoxically, the prediction that "this, too, shall pass" would then come true, not because the standards failed, but because they were never understood nor followed.

Standards are intended to demolish timeworn practices in education. Some of these practices place the teacher at the center of the classroom but reduce her or him to a cog in the machinery of the school and the district, with the primary responsibility of preparing students for tests. The standards documents recognize the need to regard teachers as professionals, students as active, independent learners, and tests as inadequate methods of assessing the full range of learning.

Within broad frameworks, the standards urge teachers to use their judgment in tailoring the curriculum to students' needs and interests. The NRC *Science Standards*, for example, call for "teachers [to be] empowered to make the decisions essential for effective learning." (1996, p. 2) Neither teachers nor administrators should interpret standards as mechanisms for tightening control over what teachers and students do. While they are very clear about the goals of education, the standards are less specific about how to meet them. Innovative curriculum efforts such as ***Stuff That Works!*** fit very well within the overall scheme of standards.

Teachers who have tried to implement ***Stuff That Works!*** activities in their classrooms have often come away with a positive feeling about them. The following comments are typical:

- *The strengths of this unit are the opportunity to group students, work on communication skills, problem solve … and plan real life tests. I have watched my students go from asking simple yes/no questions to thinking and planning careful, thoughtful active questions. The students began to see each other as people with answers... I was no longer the expert with all the answers.*
- *At second grade, with basically no prior knowledge of mechanisms, I wanted the students to start to analyze/take apart objects around them.... Most of my students really enjoyed working with mechanisms. I noticed that more girls participated in discussion than with some of my previous science activities.*
- *I must begin by telling you that I found this particular guide to be so much fun and the students demonstrated so much energy and interest in this area... I was able to engage them in the activities easily... The activities were very educational and provided so much vital information that helped students connect what is being taught to them in math to real life situations, e.g., graphing behavior and using tallies to record information. For my [special education] students, I found this gave them self confidence...*
- *I read the entire guide from front to back... Although the main idea of the unit is not specifically a large focus of instruction in our fourth grade curriculum, I recognized the power behind the*

ideas and activities and knew that this unit would promote collaboration, problem solving and communication... Overall, I think my students loved this unit and felt enormously successful after we finished...

- *My most important goal for students is that they feel good about themselves and realize what they can do. I liked these activities, because they had these results.*

The standards are intended to promote just these sorts of outcomes. When a teacher has a "gut feeling" that something is working well, there is usually some basis to this feeling. As the NRC Science Standards state, "outstanding things happen in science classrooms today... because extraordinary teachers do what needs to be done *despite conventional practice* [emphasis added]." (1996, p. 12) Unfortunately, even an extraordinary teacher may not find support from traditional administrators, who complain that the classroom is too noisy or messy, or that somebody's guidelines are not being followed. Under these circumstances, standards can be very useful. It is usually easy to see how valuable innovations fit into a national framework of education reform that is also endorsed by state- and district-level authorities. Standards can be used to justify and enhance innovative educational programs whose value is already self-evident to teachers and students.

What the Standards Really Say

In order to justify work as meeting standards, it is necessary to know what the standards really say. In the remainder of this chapter, we summarize each of the five major standards documents listed at the beginning of the chapter, and show how the *Stuff That Works!* ideas are consistent with these standards. We provide some historical background for each of the standards, and look at the overall intent and structure before relating them to the ***Stuff That Works!*** materials. These sections should be used only as they are needed. For example, if you would like to use some of the ideas from this Guide, and are also required to meet the *National Science Education Standards,* then that section could be useful to you in helping you justify your work.

Standards for Technological Literacy: Content for the Study of Technology

In April 2000, the International Technology Education Association (ITEA) unveiled the *Standards for Technological Literacy,* commonly known as the *Technology Content Standards,* after extensive reviews and revisions by the National Research Council (NRC) and the National Academy of Engineering (NAE). In its general outlines, the new standards are based on a previous position paper, *Technology for All Americans* (ITEA, 1996). The latter document defined the notion of "technological literacy" and promoted its development as the goal of technology education.

A technologically literate person is one who understands "what technology is, how it is created, and how it shapes society, and in turn is shaped by society." (ITEA, 2000, p. 9) According to the *Standards,* familiarity with these principles is important not only for those who would pursue technical careers, but for all other students as well. They will need to know about technology in order to be thoughtful practitioners in most fields, such as medicine, journalism, business, agriculture, and education. On a more general level, technological literacy is a requirement for participation in society as an intelligent consumer and an informed citizen.

Given the importance of being technologically literate, it is ironic that technology barely exists as a school subject in the U.S., and is particularly hard to find at the elementary level. In a curriculum overwhelmingly focused on standardized tests, there seems to be little room for a new subject such as technology. To make matters worse, there is considerable confusion over what the term *technology* even means. Many in education still equate it with "computers." The *Standards* advocate for technology education based on a broad definition of "technology," which is "how humans modify the world around them to meet their needs and wants, or to solve practical problems." (p. 22)

The *Technology Content Standards* describe three aspects of developing technological literacy: learning *about* technology, learning to *do* technology, and technology as a theme for curriculum integration (pp. 4-9). To learn about technology, students need to develop knowledge not only about specific technologies (Standards 14 – 20), but also about the nature of technology in general (Standards 1 – 3), including its core concepts: **systems, resources, requirements, trade-offs, processes,** and **controls.** Resources include **materials, information,** and **energy,** while **modeling** and **design** are fundamental examples of processes (p. 33). Students learn to "do" technology by engaging in a variety of technological processes, such as **troubleshooting, research, invention, problem solving, use and maintenance, assessment** of technological impact, and, of course, **design** (Standards 8 – 13). Technology has obvious and natural connections with other areas of the curriculum, including not only math and science, but also language arts, social studies, and the visual arts.

The material in the ***Stuff That Works!*** guide *Mechanisms and Other Systems* offers numerous opportunities for learning the core concepts of technology. Household mechanisms are an excellent entry point because they are familiar items that are relatively simple to analyze and understand. The operation of a mechanism is usually based on the Law of the Lever, which is an expression of the Law of Conservation of Energy. Every mechanism is an example of a system, because its parts "work together to accomplish a goal." (p. 34) A mechanism's parts are made of materials, whose properties "determine whether it is suitable for a given application." Many common mechanisms are designed to serve as tools, another category of resources, that "extend human capabilities, such as holding, lifting, carrying, fastening." (p. 35)

By analyzing how simple mechanisms work, students develop an understanding of basic systems concepts: inputs and outputs, reversibility, subsystems, and parts vs. wholes. By asking what these mechanisms do, they encounter some of the ways in which tools extend human capabilities. In the course of modeling simple mechanisms in other materials, such as cardboard, rubber bands, and paper fasteners, some of the properties of these materials become obvious. For example, it is difficult to model a spring-type mousetrap using cardboard and rubber bands because cardboard is not as stiff or strong as metal. If the rubber band is stiff enough to produce the necessary snap action, it will probably cause the cardboard to bend instead of snapping.

In the course of these modeling activities, a variety of problems arise, which become opportunities to learn troubleshooting. For example, the parts of the cardboard mechanisms get caught on one another instead of moving freely. Instead of supplying the answer, the teacher might provide a focusing question: "Where, exactly, does that piece get hung up?" Similarly, there are many opportunities for research. A popular activity involves children in creating their own cardboard mechanisms that work the same way as "mystery mechanisms," whose moving parts are concealed. As children try to solve these problems, they ask questions like: "How can I make the input and output both go the same way?" This question provides an occasion for doing some research. The teacher might suggest that they look at how the same problem has been solved by the manufacturers of some familiar mechanisms, such as a pair of tweezers, a nutcracker, or a staple remover.

Where does technology education "fit" in the existing curriculum? The Technology *Standards* address this problem by claiming that technology can enhance other disciplines: "Perhaps the most surprising message of the *Technology Content Standards* ... is the role technological studies can play in students' learning of other subjects." (p. 6) We support this claim in the following sections, which draw the connections between *Mechanisms and Other Systems* and national standards in science, math, and English language arts.

Benchmarks for Science Literacy

There are two primary standards documents for science education: The American Association for the Advancement of Science (AAAS) *Benchmarks for Science Literacy* (1993) and the National Research Council (NRC) *National Science Education Standards* (1996). Unlike the *National Science Education Standards,* the *Benchmarks* provide explicit guidance for math, technology, and social science education, as well as for science. The *Benchmarks* draw heavily on a previous AAAS report, *Science for All Americans* (1989), which is a statement of goals and general principles rather than a set of standards. *Benchmarks* recast the general principles of *Science for All Americans* (SFAA) as minimum performance objectives at each grade level.

The performance standards in *Benchmarks* are divided among 12 chapters. These include three generic chapters regarding the goals and methods of science, math and technology; six chapters providing major content objectives for the physical, life, and social sciences, technology, and mathematics; and three generic chapters dealing with the history of science, "common themes," and "habits of mind." The last four chapters of *Benchmarks* provide supporting material, such as a glossary of terms and references to relevant research.

Recognizing that standards are necessary but not sufficient for education reform, the AAAS has also developed some supplementary documents to support the process of curriculum change. These include *Resources for Science Literacy: Professional Development* (1997), which suggests reading materials for teachers, presents outlines of relevant teacher education courses, and provides comparisons between the *Benchmarks,* the Math Standards, the Science Standards and the Social Studies Standards. A subsequent publication, *Blueprints for Science Reform* (1998) offers guidance for changing the education infrastructure to support science, math, and technology education reform. The recommendations in *Blueprints* are directed towards administrators, policy makers, parent and community groups, researchers, teacher educators, and industry groups. A subsequent AAAS document, *Designs for Science Literacy* (2001), provides examples of curriculum initiatives that are based on standards.

The *Benchmarks* present a compelling argument for technology education. The authors present the current situation in stark terms: "In the United States, unlike in most developed countries in the world, technology as a subject has largely been ignored in the schools." (p. 41) Then they point out the importance of technology in children's lives, its omission from the curriculum notwithstanding: "Young children are veteran technology users by the time they enter school.... [They] are also natural explorers and inventors, and they like to make things." (p. 44) To resolve this contradiction, "School should give students many opportunities to examine the properties of materials, to use tools, and to design and build things." (p. 44)

Like the Technology Standards, the *Benchmarks* identify design as a key process of technology and advocate strongly for first-hand experience in this area. "Perhaps the best way to become familiar with the nature of engineering and design is to do some." (p. 48) As children become engaged in design, they "begin to enjoy challenges that require them to clarify a problem, generate criteria for an acceptable solution, try one out, and then make adjustments or start over again with a newly proposed solution." (p. 49) These statements strongly support the basic approach of ***Stuff That Works!***, which is to engage children in analysis and design activities based on the technologies already familiar to them. Like ***Stuff That Works!***, the *Benchmarks* also recognize the back-and-forth nature of design processes, which rarely proceed in a linear, predictable sequence from beginning to end.

In the chapter "Common Themes," *Benchmarks* identifies several "big ideas" that recur frequently in science, mathematics, and technology, and are powerful tools for explanation and design. Two of these themes, **systems** and **models**, are at least as important in technology as in science, and both are squarely addressed by work with mechanisms and circuits. The section on systems begins, "One of the essential components of higher-order thinking is the ability to think about a whole in terms of the sum of its parts and, alternatively, about parts in terms of how they relate to one another and to the whole." (p. 262) The section goes on to point out that these ideas are difficult, and learned only through studying progressively more complex examples. In grades K-2, for example, "Students should practice identifying the parts of things and see how one part connects to and affects another..." (p. 264) By the end of grade eight, they should know that "the output from one part of a system can become the input to other parts." (p. 265) A simple mechanism is an excellent place to start exploring systems concepts, because the parts and their relationships are easily identifiable, as are the inputs and outputs. Simple mechanisms can serve as prototypes for making sense of more complex systems.

Another of the common themes is models, which are "tools for learning about the things they are meant to resemble." These include mathematical and conceptual models, as well as physical models. Of these, "physical models are by far the most obvious to young children, so they should be used to introduce the idea of models." (p. 266) As children design and make their own cardboard models of mechanisms, they quickly become aware that "a model of something is different from the real thing but can be used to learn something about the real thing" (p. 268). For example, in modeling a pair of scissors in oak tag, Annette Purnell's students realized that they could make the model *move* like a real pair of scissors, but they couldn't make it cut, because it was made of different materials. (See Chapter 4, p. 117.)

The importance of learning by doing is stressed in the chapter called "Habits of Mind." The section on "Manipulation and Observation" states, "Education for science literacy implies that students be helped to develop the habit of using tools, along with scientific and mathematical ideas and computation skills, to solve practical problems..." When "things don't work right... the problem can be diagnosed and the malfunctioning device fixed using ordinary troubleshooting techniques and tools." (p. 292) Toward those ends, the document presents the following benchmarks: "By the end of second grade, students should be able to make something out of paper, cardboard, wood, plastic, metal, or existing objects, that can actually be used to perform a task." (p. 292) "By the end of eighth grade, students should be able to inspect, disassemble, and reassemble simple mechanical devices and describe what the various parts are for." (p. 294) Here are compelling reasons for engaging children with mechanisms and circuits: these activities will provide them with the experience and confidence to make, analyze, and fix things.

Work with mechanisms engages students in exploring the characteristics of materials: "Young children should have many experiences in working with different kinds of materials, identifying and composing their properties and figuring out their suitability for different purposes." (p. 188) Children develop these skills as they try to model mechanisms.

The National Science Education Standards

In 1991, the National Science Teachers Association asked the National Research Council to develop a set of national science education standards. These standards were intended to complement the *Benchmarks*, which include math, technology, and social studies as well as natural science. The National Research Council (NRC) includes the National Academy of Sciences, which is composed of the most highly regarded scientists in the country. Over the course of the next five years, the NRC involved thousands of scientists, educators, and engineers in an extensive process of creating and reviewing drafts of the new science standards. The results were published in 1996 as the *National Science Education Standards* (NSES).

Who is the audience for standards? The conventional view is that standards outline what students should know and be able to do, and that teachers are accountable for assuring that their students meet these guidelines. The NSES take a much broader approach, looking at the whole range of systemic changes needed to reform science education. The document is organized into six sets of standards. Only one of the six, the "Science Content Standards," talks directly about what children should learn through science education. The other five address other components of the education infrastructure, including classroom environments, teaching methods, assessment, professional development, administrative support at the school and district levels, and policy at the local, state, and national levels.

Collectively, these standards outline the roles of a large group of people on whom science education depends: teachers, teacher educators, staff developers, curriculum developers, designers of assessments, administrators, superintendents, school board members, politicians, informed citizens, and leaders of professional associations. If an administrator or school board member were to ask a teacher, "What are you doing to address the *National Science Education Standards*?" the teacher would be fully justified in responding, "What are *you* doing to meet them?"

One message that recurs frequently in the NSES is that teachers must be regarded as professionals, with a vital stake in the improvement of science education and an active role "in the ongoing planning and development of the school science program." (p. 50) More specifically, they should "participate in decisions concerning the allocation of time and other resources to the science program." (p. 51) The Standards explicitly reject the reduction of teachers to technicians or functionaries who carry out somebody else's directives. "Teachers must be acknowledged and treated as professionals whose work requires understanding and ability." The organization of schools must change too: "School leaders must structure and sustain suitable support systems for the work that teachers do." (p. 223)

Teachers should also play a major role in deciding and/or designing the science curriculum. The *Standards* call for teachers to "select science content and adapt and design curricula to meet the needs, interests, abilities and experiences of students." Although teachers set the curriculum initially, they should remain flexible: "Teaching for understanding requires responsiveness to students, so activities and strategies are continuously adapted and refined to address topics arising from student inquiries and experiences, as well as school, community and national events." (p. 30) Not only teachers, but also students, should play a major role in curriculum planning. The Teaching Standards make this point explicit: "Teachers [should] give students the opportunity to participate in setting goals, planning activities, assessing work and designing the environment." (p. 50)

More specifically, Content Standard E, "Science and Technology," strongly supports the approach of ***Stuff That Works!:*** "Children's abilities in technological problem solving can be developed by firsthand experience in tackling tasks with a technological purpose. They can also study technological products and systems in their world—zippers, coat hooks, and can openers… They can study existing products to determine function and try to identify problems solved, materials used and how well a product does what it is supposed to do… Tasks should be conducted within immediately familiar contexts of the home and school." (p. 135)

The Science Standards do not make the distinction between design and inquiry as sharply as do the Technology Standards: "Children in grades K-4 understand and can carry out design activities earlier than they can inquiry activities, but they cannot easily tell the difference between the two, nor is it important whether they can." (p.135) Thus, many of the abilities and concepts needed to meet the standard "Science as Inquiry" are also developed through design. These include: "Ask a question about objects… in the environment"; "plan and conduct a simple investigation"; "employ simple equipment and tools to gather data"; and "communicate investigations or explanations." (p. 122)

The material in *Mechanisms and Other Systems* is of particular relevance to the K-12 Content Standards, "Unifying Concepts and Processes." Two of the five unifying themes are "systems, order and organization" and "form and function." As already mentioned, either a mechanism or a circuit can serve as a prototype for developing the concept of a system, which the NSES defines as "an organized group of related components or objects that form a whole." (p. 116) Similarly, examination of a mechanism, and close analysis of how it works, can be a prototype for learning that "the form or shape of an object or system is frequently related to use, operation or function." (p. 119)

The material in *Mechanisms and Other Systems* also addresses Content Standard B, "Physical Science," for grades K-4. Sorting and classifying mechanisms, for example, is an obvious way to discover the "properties of objects and materials" and the "similarities and differences of the objects." As they look at how simple mechanisms operate, children develop their own understanding of the "position and movement of objects." (p. 125) Their work with electric switches and other control devices helps them "begin to understand that phenomena can be observed, measured and controlled in various ways." (p. 126)

Principles and Standards for School Mathematics

The first of the major standards documents, *Curriculum and Evaluation Standards for School Mathematics*, was published in 1989 by the National Council of Teachers of Mathematics (NCTM). Additional standards for teaching and assessment were published in 1991 and 1995, respectively. In 2000, the NCTM released a new document, *Principles and Standards for School Mathematics*, intended to update and consolidate the classroom-related portions of the three previous documents. Some of the major features of the new volume, different from the prior version, are the addition of the Principles, the division of the standards into the categories "Content" and "Process," and the inclusion of a new process standard called "Representation."

The new NCTM document acknowledges the limitations of educational standards: "Sometimes the changes made in the name of standards have been superficial or incomplete... Efforts to move in the direction of the original NCTM Standards are by no means fully developed or firmly in place." (pp. 5-6) In spite of this candid assessment, the authors remain optimistic about the future impact of standards. Their goal is to provide a common framework for curriculum developers and teachers nationwide. If all schools follow the same standards, then teachers will be able to assume that "students will reach certain levels of conceptual understanding and procedural fluency by certain points in the curriculum." (p. 7)

The NCTM *Principles and Standards* begin by presenting the six sets of principles, which are the underlying assumptions for the standards. Some of these principles are common to the other standards documents: that there should be high expectations of all students, that the goal of learning is deep understanding, and that assessment should be integrated with curriculum. Other principles underscore the need to learn from cognitive research. More than in any other field, there has been extensive research into how students learn mathematics, and this research base is reflected in the *Principles*. For example, the "Curriculum Principle" calls for coherent sets of lessons, focused collectively on one "big idea." Similarly, the "Teaching Principle" specifies that teachers must be aware of students' cognitive development. The "Learning Principle" cites research on how learning can be most effective.

The standards themselves are organized into two categories: Content Standards and Process Standards. The former describe what students should learn, in the areas of Number and Operations, Algebra, Geometry, Measurement, and Data Analysis and Probability. The Process Standards discuss how students should acquire and make use of the content knowledge. The subcategories are Problem Solving, Reasoning and Proof, Communication, Connections, and Representation. Unlike the earlier NCTM document, the new version uses all the same standards across all of the grade levels, from K through 12. In this way, the NCTM is advocating for a carefully structured curriculum, which builds upon and extends a few fundamental ideas systematically across the grades. Readers may be surprised to find an Algebra Standard for grades K-2, or a Number and Operations Standard for grades 9-12.

Stuff That Works! units and activities offer rich opportunities for fulfilling a key ingredient of the NCTM standards: learning and using mathematics in context. The Process Standard called "Connections" makes it clear that mathematics should be learned by using it to solve problems arising from "other subject areas and disciplines" as well as from students' daily lives."

(p. 66) ***Stuff That Works!*** fulfills this standard in two fundamental respects: it provides mathematics connections with another subject area, technology, and it uses artifacts and issues from everyday life as the source of material for study. The mathematics students learn is drawn from all of the Content Standards, as well as all of the Process Standards except for Reasoning and Proof.

Mechanisms and Other Systems engages students in basic learning about spatial relationships. Students sketch simple mechanisms in their open and closed positions, or try to draw the basic topographical relationships of simple circuits. Often, these representations become a form of communication about how to make things or about how they work. Modeling a mechanism is an engaging activity that requires close attention to geometric relationships and, often, measurement as well. If the model does not work the same way as the original mechanism, it may indicate that the parts are not of the proper shapes or the correct proportions, and that calls for problem solving.

Sorting and classifying everyday objects are very popular starting activities in *Mechanisms and Other Systems.* These activities prepare the way for the more formal methods of pattern handling known as algebra. The NCTM strongly recommends that these basic ideas about patterns be developed with very young children. The Algebra Standard for grades K-2 calls for pattern finding and pattern recognizing activities, such as classifying and sorting, and identifying "the criteria [students] are using as they sort and group objects." Basic classifying activities are designed to "help students develop the ability to form generalizations." (p. 91) As part of *Mechanisms and Other Systems,* students sort mechanisms, switches, and other controls, and ask other students to "guess what our categories were" just by looking at the objects in each group.

Standards for the English Language Arts

By 1991, it had become clear that standards would be produced for all of the major school subjects. Fearful that English language standards might be produced without a firm basis in research and practice, two major professional organizations requested Federal funding for their own standards effort. The following year, the Department of Education awarded a grant for this purpose to the Center for the Study of Reading at the University of Illinois, which agreed to work closely with the two organizations, the National Council of Teachers of English (NCTE) and the International Reading Association (IRA). This effort culminated in the 1996 publication of the *Standards for the English Language Arts* by the NCTE and IRA. These ELA Standards are now widely accepted for their clear, concise outline of English language education.

The ELA *Standards* adopt an unusually comprehensive view of "literacy," much broader in its scope than the traditional definition of "knowing how to read and write." (p. 4) Literacy also includes the ability to think critically, and encompasses oral and visual, as well as written communication. Recognizing that these forms of language "are often given limited attention in the curriculum," the *Standards* outline the variety of means used to convey messages in contemporary society:

> "Being literate in contemporary society means being active, critical, and creative users not only of print and spoken language, but also of the visual language of film and television, commercial and political advertising, photography, and more. Teaching students how to interpret and create visual texts such as illustrations, charts, graphs,

electronic displays, photographs, film and video is another essential component of the English language arts curriculum." (pp. 5-6)

According to the ELA *Standards*, there are three major aspects to language learning: **content**, **purpose**, and **development**. Content standards address only *what* students should learn, but not why or how: "Knowledge alone is of little value if one has no need to – or cannot – apply it." The *Standards* identify four purposes for learning and using language: "for obtaining and communicating information, for literary response and expression, for learning and reflection, and for problem solving and application." (p.16) Purpose also figures prominently in the third dimension of language learning, development, which describes *how* students acquire this facility. "We learn language not simply for the sake of learning language; we learn it to make sense of the world around us and to communicate our understanding with others." (p. 19)

Of course, purpose and motivation vary from one situation to another. The authors of the *Standards* make this point, too, in their discussion of "context." "Perhaps the most obvious way in which language is social is that it almost always relates to others, either directly or indirectly: we speak to others, listen to others, write to others, read what others have written, make visual representations to others and interpret their visual representations." Language development proceeds through the practice of these communication skills with others: "We become participants in an increasing number of language groups that necessarily influence the ways in which we speak, write and represent." While language development is primarily social, there is an individual dimension as well: "All of us draw on our own sets of experiences and strategies as we use language to construct meaning from what we read, write, hear, say, observe, and represent." (p. 22)

How does this broad conception of literacy and its development relate to daily classroom practice? The authors recognize that the ELA *Standards* may be in conflict with the day-to-day demands placed on teachers. "They may be told they should respond to the need for reforms and innovations while at the same time being discouraged from making their instructional practices look too different from those of the past." Among those traditional practices are the use of standardized tests, "which often focus on isolated skills rather than contextualized learning." Prescribed texts and rigid lesson plans are further barriers to reform, because they tend to preclude "using materials that take advantage of students' interests and needs" and replace "authentic, open-ended learning experiences." (p. 7) Another problem is "the widespread practice of dividing the class day into separate periods [which] precludes integration among the English language arts and other subject areas." (p. 8) Taken seriously, these standards would lead to wholesale reorganization of most school experiences.

This introductory material sets the stage for the twelve content standards, which define "what students should know and be able to do in the English language arts." (p. 24) Although these are labeled "content" standards, "content cannot be separated from the purpose, development and context of language learning." (p. 24) In a variety of ways, the twelve standards emphasize the need to engage students in using language clearly, critically and creatively, as participants in "literacy communities." Within these communities, students sometimes participate as *receivers* of language—by interpreting graphics, reading and listening and—and sometimes as *creators*—by using visual language, writing, and speaking.

Some teachers have used the ***Stuff That Works!*** activities and units primarily to promote language literacy, rather than for their connections with math or science. Technology activities offer compelling reasons for children to communicate their ideas in written, spoken, and visual form. In early childhood and special education classrooms, teachers have used ***Stuff That Works!*** to help children overcome difficulties in reading and writing, because it provides natural and non-threatening entry points for written expression. In the upper elementary grades, ***Stuff That Works!*** activities offer rich opportunities for students to want to use language for social purposes. Several characteristics of ***Stuff That Works!*** contribute to its enormous potential for language learning and use:

- Every unit begins with an extensive group discussion of what terms mean, how they apply to particular examples, how to categorize things, and/or what problems are most important.
- The activities focus on artifacts and problems that engage children's imaginations, making it easy to communicate about them. Teachers who use ***Stuff That Works!*** usually require students to record their activities and reflections in journals.

For each of the ***Stuff That Works!*** topics, the opening activity is a scavenger hunt or brainstorming session.

In a brainstorming session, students think of the examples, list them, and then try to make sense of them. Often, the teacher starts the discussion by asking the students to define a word. These discussions can be rich opportunities to explore and inquire about language.

For example, at the beginning of a unit on mechanisms, Mary Flores asked her special education students, "What is a mechanism?" One of Mary's students said, "I see the word *mechanic.*" Another said that it sounds like the word *mechanical.* A student from another group said, "We are mechanisms because we move around." (See Chapter 4, p. 128.) The students in this class were "draw[ing] on their prior experience, their interactions with other readers and writers [and] their knowledge of word meaning and of other texts," to make sense of this new word (ELA Standard #3, p. 31).

Many special education students have very low self-esteem and are deeply frustrated by the difficulties they experience in learning to read and write. By focusing on tangible and interesting artifacts, ***Stuff That Works!*** activities offer unusual opportunities to overcome some of these barriers. Faced with a group who were unwilling to write at all, Kathy Aguiar asked them to talk about and draw pictures of various mechanical devices, and to make a one-word label showing the "control." Before long, they were writing full sentences describing how they thought these devices worked. (See Chapter 4, p. 121.) These students were learning "to adjust their use of spoken, written and visual language to communicate effectively." (ELA Standard #4, p. 33)

GLOSSARY

Analog control: A control device that uses a continuously variable quantity to select the desired outcome; for example, a volume control on a radio, or a dimmer control for a light.

Atom: The smallest unit of matter that preserves the chemical characteristics of a material. Atoms are electrically neutral, because the positively-charged protons in the nucleus are surrounded by the same number of negatively-charged electrons.

Automatic control system: A control system that supplies its own control input by sampling its own output; for example, a thermostat-controlled home heating system.

Charge: A discrete quantity of electricity.

Circuit: The complete path of an electric current including the source of electric energy.

Closed loop system: Another term for "automatic control system." The closed loop is the path from the output back to the input, and is also known as the "feedback loop."

Closed system: A system that conserves its own material and does not depend on inputs from the outside. For example, a steam heating system recirculates the same water over and over again.

Compound lever: A device in which one lever acts on another.

Compression spring: A spring whose ends try to push apart when they are compressed.

Conductor: A material that permits an electric current to flow easily.

Control: A device that uses a negligible amount of energy to affect a much larger energy flow.

Control system: A system that uses a control input to manage an energy flow.

Current: A flow of electrical charge.

Digital control: A control device that uses a preset number of possible states to select the desired outcome; for example, an ON/OFF switch, or the buttons for selecting the stations on a car radio.

Effort: The point on a lever where force is applied in order to exert a force at another point; also, the amount of force applied to the effort point.

Effort arm: The distance on a lever from the effort to the fulcrum.

Electron: A sub-atomic particle that carries most of the current in electrical circuits.

Feedback loop: A connection from the output to the input of an automatic control system, which provides control information based on the current status of the system.

First-class lever: A lever arranged with the fulcrum between the effort and the load.

Fixed pivot: A pivot that attaches a link to the base of a mechanism.

Floating pivot: A pivot that attaches two links to each other, but does not fix them to the base of the mechanism.

Flow: Movement of a fluid or electric charge from one point to another

Fulcrum: The support or pivot around which a lever turns.

Gear: A toothed wheel designed to turn another gear with similar teeth.

Inclined plane: A ramp used to lessen the amount of force needed by increasing the distance over which a load must travel.

Input: A point where energy, material, or information is introduced into a system.

Insulator: Material that is a poor conductor of electricity or heat.

Ion: An atom that has lost or gained one or more electrons, giving it a net positive or negative charge.

Joint: A connection between two links or between a link and the base. Major joint types are the pivot, slider, and roll-slide joint.

Law of the Lever: In a lever, the force ratio of load to effort is equal to the ratio of load arm to effort arm. This ratio is called the "mechanical advantage."

Lever: A rigid bar, turning on a fulcrum, used to exert a force at one point along its length, called the "load," by the application of force at a second point, known as the "effort."

Linear motion: Motion along a straight line.

Link: A rod or bar that transmits force and motion to or from other links through joints.

Linkage: A system of links connected by joints.

Load: The output force that results from applying a force at the effort point on a lever; also, the point on the lever where the load force acts.

Load arm: The distance on a lever from the load to the fulcrum.

Manual control system: A control system in which a human user operates the control input; also known as an "open lop system."

Mechanical advantage: The ratio of load to effort force of a lever; also the ratio of effort arm to load arm. The fact that these two are equal is the Law of the Lever.

Mechanism: A device with moving parts that converts force and motion at one point to a different combination of force and motion at another point. A linkage is an example of a mechanism.

Neutral: Having an equal number of positive and negative charges.

Open loop system: A manual control system; so called because there is no feedback loop connecting the output to the input.

Open system: A system that does not conserve material, and therefore requires new inputs from the outside.

Oscillating motion: Back-and-forth motion along a circular path; for example, the motion of an agitator in a washing machine.

Output: The motion that results when energy is put into a system.

Parallel connection: A side-by-side arrangement of switches or other devices in an electric circuit so that the current can travel along either or both of two or more paths.

Pivot: A shaft or pin on which something turns; often used interchangeably with "fulcrum."

Pulley: A device that transmits force and motion by means of a rope, string, or cable wrapped around a grooved wheel.

Reciprocating motion: Back-and-forth motion in a straight line.

Return spring: A spring whose purpose is to return a mechanism to its original position when the input is released.

Roll-slide joint: A joint that combines a slider and a pivot, allowing both translation and rotation.

Rotary motion: Motion that follows a circular path.

Rotation: Clockwise or counterclockwise rotary motion.

Schematic diagram: A map showing the components and connections of an electric circuit.

Screw: A simple machine that consists of an inclined plane wrapped around a cylinder or cone.

Second-class lever: A lever arranged so that the load is between the fulcrum and the effort.

Sensor: A monitoring device that provides information about the output to the control input of an automatic control system.

Series connection: A sequential arrangement of switches or other devices in an electric circuit so that all current flowing through one must also flow through the others.

Simple machine: One of several elementary machines once considered to be the elements of which all machines are composed: the lever, the wheel-and-axle, the pulley, the inclined plane, the wedge, and the screw.

Slider: A joint that allows one link to move in a straight line with respect to another link or the base.

Subsystem: A system that is part of a larger system.

Switch: A device for making or breaking a connection in an electrical circuit.

System: A collection of interconnected parts functioning together in a way that makes the whole greater than the sum of its parts.

Technology: The artifacts, systems, and environments designed by people to improve their lives.

Tension spring: A spring whose ends try to pull back together when they are extended.

Terminal: The part of an electrical device that is used to make connections in an electrical circuit.

Third-class lever: A lever arranged so that the effort is between the fulcrum and the load.

Tool: A device that aids in accomplishing a task.

Translation: Linear motion in one direction.

Two-dimensional linkage: A linkage in which all links have to move within the same plane surface.

Wedge: A triangular device with an inclined plane on either side.

Wheel-and-axle: A large diameter cylinder turned by or turning a long circular rod.

REFERENCES

Chapter 2

Ardley, Neil. (1992) ***The Science Book of Machines.*** New York: Harcourt Brace Jovanovich.

This book teaches elementary-age children to make a variety of mechanisms from everyday materials. Included are a gear-driven fan and a toy that changes a rotary-crank-driven input into a "jumping jack" reciprocating output.

Bailey, Omry, et al. (1981) ***Modular Courses in Technology: Mechanisms.*** Edinburgh, UK: Oliver & Boyd.

Written at Junior High School level, this book teaches basic concepts of mechanisms with diagrams and photographs, mainly of Fischertechnik components. Most of the focus is on gear, pulley and cam mechanisms. Linkages are handled only briefly at the end.

Bolt, Brian. (1991) ***Mathematics Meets Technology.*** New York: Cambridge University Press.

Aimed at high school teachers of math and technology, this book is designed to show how mechanisms can provide a basis for exploring elementary algebra and geometry. It is also the only reference we have found which discusses the adjustable triangle and four-bar linkages at an elementary level, and contains a wealth of ideas for exploring and designing these mechanisms.

Bridgman, George B. (1966) ***The Human Machine: the Anatomical Structure and Mechanism of the Human Body.*** New York: Barnes & Noble.

Intended for artists, this book uses mechanical analogies to explain the joints and links that account for movement in our bodies. The drawings are very detailed and insightful.

Bureau of Naval Personnel. (1971) ***Basic Machines and How They Work.*** New York: Dover Publications.

Intended as a training manual for sailors, this book provides a fundamental background in mechanisms. As one would expect, many of the examples come from shipboard equipment, such as hoists, winches and watertight hatches. However, there is also an excellent and very well illustrated explanation of the manual typewriter in Chapter 12.

Bureau of Naval Personnel. (1973) ***Tools and Their Uses.*** New York: Dover Publications.

This is a very readable, brief introduction to all of the common hand and power tools. Many of these tools are themselves mechanisms, such as adjustable wrenches, clamps and vise-grip pliers.

Catlin, Dave. (undated) ***The Inventa Book of Mechanisms.*** London: Valiant Technology, Ltd.

Aimed at upper elementary and junior high school students, this is mostly a cartoon book that describes the simple machines in a very entertaining fashion. For example, gear ratios are discussed by talking gears who are engaged in an argument over whether large or small gears are better.

De Bono, Edward. (1972) ***Children Solve Problems.*** London, UK: Penguin Books.

De Bono, Edward. (1970) ***The Dog Exercising Machine.*** London, UK: Jonathan Cape, Ltd.

De Bono is a well-known writer of books on techniques for improving one's thinking skills. These two books look at thought processes from a somewhat different perspective. They document several fascinating studies in which children were asked to produce concept designs for solving some everyday problems. Some of these lent themselves to mechanical solutions: Figure out a way to stop a dog and cat from fighting. Invent a bicycle for postmen.

Driver, R., Squires, A., Rushworth, P. & Wood-Robinson, V. (1994) ***Making Sense of Secondary Science: Research Into Children's Ideas.*** London: Routledge.

This book is an authoritative compendium of current research on the conceptions actually held by children about physical and biological phenomena. There are chapters on electricty, forces and motion.

Dunn, Andrew. (1993) ***Lifting by Levers.*** New York: Thomson Learning.

Suitable for third and fourth grades, this book shows how to find levers in a variety of tools, utensils, toys, sports equipment and other common objects. It is well written and well illustrated.

Dunn, Andrew. (1992) ***Wheels at Work.*** New York: Thomson Learning.

A children's book, aimed at the middle elementary grades, which reveals how wheels and axles can be found in the most unlikely places, and makes the analogy between wheels and levers.

Dunn, Susan & Larson, Robb. (1990) ***Design Technology: Children's Engineering.*** Philadelphia: The Falmer Press.

Using many examples from both the UK and the US, this book develops a pedagogical basis for design technology in the elementary grades. The beautiful illustrations include photographs of children at work, as well as pictures and sketches of their designs.

Erdman, Arthur & Sandor, George. (1997) ***Mechanism Design, Vol. 1: Analysis and Synthesis*** (Third Edition). Upper Saddle River, NJ: Prentice Hall.

Though written in technical language, the first two chapters illustrate the subject of mechanisms with an extraordinary wealth of drawings, photos and diagrams of real mechanisms. A CD is included which contains animated displays and background information describing many of these mechanisms.

Eyewitness Visual Dictionaries. (1991) ***The Visual Dictionary of Everyday Things.*** New York: Doring Kindersley.

This entire book consists of photographs of disassembled objects, with every piece labeled. No explanations are given of how these pieces fit together, let alone how they work, but children who like to take things apart could use this book as a reference. The umbrella, the lamp and the bicycle are among the most interesting.

Graf, Rudolf F. & Whalen, George J. (1982) ***How It Works Illustrated: Everyday Devices and Mechanisms.*** London: Souvenir Press.

This Popular Science book provides two pages each to explain nearly one hundred common mechanisms, including some not found in other books. The explanations are generally clear and well illustrated. Some of the best are the rotary telephone dial, the tire pressure gauge, the vise-grip pliers, the electric can opener, the traffic light, the bathroom scale, the staple gun and the Medeco lock.

Guilbaud, G.T. (1960) ***What Is Cybernetics?*** New York: Grove Press.

One of the few non-technical introductions to the field of cybernetics, which is the study of automatic control systems.

Hassenstein, B. (1971) ***Information and Control in the Living Organism.*** London: Chapman and Hall.

A brief, largely non-technical book with many examples of feedback in nature, particularly in the human body.

Hiner, Mark. (1985) ***Paper Engineering for Pop-up Books and Cards.*** Sheffield, UK: Tarquin Publications.

This book has cutout pages and detailed directions for making ten mechanisms out of paper, including pull-tab linkages and pop-up parallelograms. The most valuable aspect of the book is the page of "Technical considerations" accompanying each mechanism. These pages discuss the principles involved in each case, and offer suggestions for fine-tuning the design.

Jackson, Paul. (1996) ***The Pop-Up Book.*** New York: Henry Holt & Co.

This lavishly illustrated book provides detailed instructions for making more than 100 different kinds of pop-up mechanisms, ranging from the simple to the elaborately complex. It includes many ideas for experimentation, as well as photos of professional finished products.

Johnson, Paul. (1994) ***Pop-Up Paper Engineering: Cross-Curricular Activities in Design Technology, English and Art.*** Philadelphia: The Falmer Press.

This is a detailed manual, intended for teachers, which incorporates a how-to guide for making pop-ups, brief explanations of the principles involved and suggestions for integration with art and language.

Kelly, John, et al. (1994) ***The Robot Zoo: A Mechanical Guide to the Way Animals Work.*** Atlanta: Turner Publishing, Inc.

This lavishly illustrated book shows how animals operate by replacing the animals' body parts with machine components. The resulting drawings have a robot-like, sci-fi appearance. For examples, the grasshopper's legs have springs in them. The glossary gives a brief discussion of basic animal physiology.

Kempton, W. (1987) "Two Theories of Home Heat Control." In Holland, D. & Quinn, N. (eds.). ***Cultural Models in Language and Thought.*** Cambridge: Cambridge University Press, Pp. 222-242

A fascinating account of how adults think a thermostat works.

Lehrer, Richard & Schauble, Leona. (1998) "Reasoning about Structure and Function: Children's Conceptions of Gears," in ***Journal of Research in Science Teaching,*** 35:1, January 1998, Pp. 3-25.

This is one of the few contemporary studies of children's cognitive development in analysis of mechanisms. The authors studied how second- and fifth-graders explain the operation of an egg beater, a ten-speed bike and various configurations of meshed gears on a gear board. The paper demonstrates clearly how children's conceptions evolve; for example, many children believed that the speed depends on the number of gears!

Macaulay, David. (1998) ***The New Way Things Work.*** Boston: Houghton Mifflin Co.

This newly revised classic is indispensable for beginning a study of mechanisms, and inspired much of the work described here. It has beautiful drawings showing the operation of hundreds of everyday devices in exquisite detail, including the two-arm corkscrew, the automotive window winding mechanism, the sewing machine, the bicycle brake, the toilet tank, the basic circuit, the battery and many, many, more. The book is written at a non-technical adult level, but the drawings are useful for all levels.

Marzio, Peter. (1973) ***Rube Goldberg: His Life and Work.*** New York: Harper & Row.

This comprehensive biography of Rube Goldberg includes many insights into the messages inherent in his fantastic creations, and also contains many of his original drawings.

McDermott, Lillian C. & Schaffer, Peter S. (1992) "Research as a Guide for Curriculum Development: An example from introductory electricity," in ***American Journal of Physics,*** 60:11, Pp. 994-1013.

The first part of this article presents research on how college students understand electricity, after being taught the scientific view in a college physics course. Their conceptions are remarkably unaffected by traditional classroom instruction! In the second part, the authors present instructional methods that could challenge the students' prior conceptions directly.

Myska, David. (1999) ***Machines and Mechanisms: Applied Kinematic Analysis.*** Upper Saddle River, NJ: Prentice Hall.

This book is written for two-year college students of engineering technology. Like Norton (1999), much of the book is difficult mathematically, but the first chapter provides a clear introduction to the subject, and there are many drawings showing examples of mechanisms, particularly in the problems at the ends of the chapters.

Norton, Robert L. (1999) ***Design of Machinery: An Introduction to the Synthesis and Analysis of Mechanisms and Machines*** (Second Edition). New York: McGraw-Hill.

This is a college-level textbook in mechanism design for engineering students. Although most of the book is mathematically demanding, the first two chapters provide a clear introduction to basic concepts, with many examples from everyday life.

Odum, Eugene P. (1975) ***Ecology: The Link Between the Natural and the Social Sciences.*** New York: Holt, Rinehart and Winston.

A brief, introductory college-level book on ecology, which makes extensive use of feedback concepts.

Penner, David, et al. (1997) "Building Functional Models: Designing an Elbow," in *Journal of Research in Science Teaching,* 34:2, February 1997, Pp. 125-143.

This is a report of a study of first- and second-graders modeling a familiar mechanism, the human elbow. It includes two cycles of redesign, as the children improved their models to make them function more like a real elbow. In addition to the excellent examples of design and re-design, the paper shows how children understand the concept of a model.

Petroski, Henry. (1994) ***The Evolution of Useful Things.*** New York: Random House.

Petroski's primary argument is that designs evolve based on the shortcomings and outright failures of their predecessors. He makes this point through an exhaustive account of how such devices as the hammer, the can opener and the paper clip attained their present form. This book provides a useful historical background for discussing a variety of everyday technologies.

Piaget, Jean. (1978) ***Success and Understanding.*** Cambridge, MA: Harvard University Press.

In great detail, Piaget explores the development of children's understanding of some basic artifacts, including levers, seesaws, toy cars and bridges. As in so much of Piaget's work, the step-by-step approach reveals not only how children's ideas-take form, but also provides invaluable insights into what the concepts are and how they are connected.

Piaget, Jean. (1972) ***The Child's Conception of Physical Causality.*** Totowa, NJ: Littlefield, Adams & Co.

This is one of Piaget's classic works, which explores in depth how children at various levels understand the causes of common phenomena. Chapter 9, "The Mechanism of Bicycles," uncovers children's understanding of the motion of the pedals and of the rear wheels, and gives examples of fantastic alternative conceptions. The concluding chapter is a very valuable summary of how children's beliefs about causality evolve.

Reader's Digest. (1983) ***Complete Do-It-Yourself Manual.*** Pleasantville, NY: Reader's Digest Associates.

This comprehensive manual of home repair contains a wealth of information on mechanisms and circuits, including hand and power tools, drapery hardware, garage door openers, toilets, faucets, locks, lighting fixtures and window hardware. The illustrations are excellent.

Scrine, Rick & Clewes, Eric. (1989) ***First Technology.*** London: Hodder and Stoughton.

Intended for Junior High School level, this book has a long section called "Control Systems" which deals mostly with mechanisms. It offers a clearly written introduction to machines, and a few design problems, using mostly Lego and Meccano technologies.

Shooter, Ken & Saxton, John. (1997) ***Making Things Work: An Introduction to Design Technology.*** New York: Cambridge University Press.

Intended for use at Junior High School level, this entertaining book deals with electronics and structures as well as mechanisms. Many open-ended design and inquiry activities are suggested.

Staffordshire Design and Technology Education Programme. (1992) ***S.T.E.P. Design and Technology 5-16, Key Stage 2 Datafile.*** New York, Cambridge University Press.

This package contains 126 photocopyable cards dealing with the operation and design of a wide range of technologies.

Many of the cards offer useful suggestions about how to make and test mechanisms inexpensively. See, for example, "Card and paper," "Gears," "Linkages," "Movement" and "Pulleys."

Tasker, Ross & Osborne, Roger. (1985) "Science Teaching and Science Learning," in ***Learning in Science: The Implications of Children's Science.*** Auckland, NZ: Heinemann, Pp. 15-27.

This is a thoughtful account of how children understand electric circuits. It presents the "Single-wire", "Clash of Currents" and "used-up current" theories in their clearest forms. It also shows how traditional instruction ignores these theories.

Van Cleave, Janice. (1993) ***Machines: Mind-Boggling Experiments You Can Turn into Science Fair Projects.*** New York: John Wiley & Sons.

This book describes twenty mechanism investigations, all of them made with free or very inexpensive materials. Included are directions for making gears from clay and toothpicks, a windlass from a broom, and a linkage that reduces or enlarges the scale of drawings. It is written at upper elementary grade level.

Walton, Harry. (1968) ***The How and Why of Mechanical Movements.*** New York: Popular Science Publishing Company.

Written in an entertaining, accessible style, this book offers clear explanations of the lever, wheel, inclined plane, screw, pulley and gear. The examples are well illustrated, and range from simple technologies like the hammer and water wheel to automatic transmissions, jet engines and aircraft guidance systems.

Williams, Pat & Jinks, David. (1991) ***Design and Technology 5-12.*** Philadelphia: The Falmer Press.

Based on many experiences in the UK, this book makes a convincing argument for teaching design technology in the elementary grades. It also provides an outline of basic principles, both of content and pedagogy, and numerous case studies of curriculum units. "Levers and Ducks" describes a unit in which six-year-olds designed and made toy paper ducks with flapping wings controlled by linkages much like those described in this chapter.

Zubrowski, Bernie. (1993) ***Mobiles: Building and Experimenting with Balancing Toys.*** New York: Beech Tree Books.

In his usual entertaining style, Zubrowski outlines a wealth of activities designed to help children explore what makes things balance. By the end of the book, the mobiles are unsymmetrical and quite complex.

Zubrowski, Bernie. (1986) ***Wheels at Work: Building and Experimenting with Models of Machines.*** New York: Beech Tree Books.

In this book, Zubrowski provides detailed instructions for making a block and tackle from clothesline pulleys; gears, waterwheels and windmills using cans, plastic cups and plates, and other cheap materials; and other related contraptions. Each set of activities includes a "What's happening" section, which discusses physics concepts and "Experiments to try" which proposes areas for investigation. The material is most suitable for upper elementary grades and middle school, and is very well thought out.

Chapter 6

American Association for the Advancement of Science. (1989) *Science for All Americans: A Project 2061 Report on Literacy Goals in Science, Mathematics and Technology.* Washington, DC: Author.

American Association for the Advancement of Science. (1993) *Benchmarks for Science Literacy.* New York: Oxford University Press.

American Association for the Advancement of Science. (1997) *Resources for Science Literacy.* New York: Oxford University Press.

American Association for the Advancement of Science. (1998) *Blueprints for Reform.* New York: Oxford University Press.

American Association for the Advancement of Science. (2001) *Designs for Science Literacy.* New York: Oxford University Press.

International Technology Education Association. (1996) *Technology for All Americans: A Rationale and Structure or the Study of Technology. Reston,* VA: Author.

International Technology Education Association. (2000) *Standards for Technological Literacy: Content for the Study of Technology.* Reston, VA: Author.

National Center on Education and the Economy. (1997) *New Standards Performance Standards; Vol 1: Elementary School.* Washington, DC: Author

National Council for the Social Studies. (1994) *Expectations of Excellence: Curriculum Standards for Social Studies.* Washington, DC: Author.

National Council for the Social Studies. (1997) *National Standards for Social Studies Teachers.* http://www.socialstudies.org/standards/teachers/standards.html

National Council of Teachers of English & International Reading Association. (1996) *Standards for the English Language Arts.* Urbana, IL: Author.

National Council of Teachers of Mathematics. (1989) *Curriculum and Evaluation Standards for School Mathematics.* Reston, VA: Author.

National Council of Teachers of Mathematics. (2000) *Principles and Standards for School Mathematics.* Reston, VA: Author.

National Research Council. (1996) *National Science Education Standards.* Washington, DC: National Academy Press.